The Joy of Automobile Repair

The Joy of Automobile Repair

by C. J. Becker

Drawings by Jennifer Ladner

LANCASTER-MILLER PUBLISHERS
Berkeley, California
1979

3165 Adeline St.,
Berkeley, California 95703
Printed and bound in the United States of America

ISBN 0-89581-025-5

Cover design Brenton Beck, Fifth Street Design Associates

CONTENTS

Chapter 4: Fuel Delivery I: Fuel Pump

Chapter 5: Fuel Delivery II: Fuel Injection

Chapter 13: Brakes, Axles, and Bearings

Chapter 14: The Clutch: Driveshaft and U-Joints

Chapter 15: Engine Overhaul (Bottom End)

Chapter 16: Standard Transmission and Differential

For Jackie and Cyrus

Preface

During all the years I have worked on cars, this is the book I needed but could never find. So now I have written it myself.

It is a manual for the practicing auto mechanic, the advanced amateur, or anybody who loves (hates) the automobile and can read. It does not tell how to change your radiator water with the garden hose, or how often to rotate your tires, or how to set the points with a match-book cover while you are stranded on the Mojave Desert.

What it does include is advice on the repair problems in the modern automobile, especially in areas where little or no information is available—rebuilding and tuning SU carburetors; trouble shooting electronic injection and ignition with inexpensive equipment; rebuilding overhead-cam heads and wet-sleeve cylinder blocks; discussion of practical aspects of timing, more complete than any I know in print, starting with timing the crank to the cam from scratch to timing the distributor by experiment, and consideration of most of the more conventional repair topics.

The material is arranged according to familiar repair topics (the Brake, the Clutch) and in cumulative self-teaching order. Various approaches to each rebuild problem are presented, if appropriate, with the probability for success. Many diagnostic routines are unique in this book. Others go back at least as far as *Dyke's Automobile and Gasoline Engine Encyclopedia* (1918).

Among this book's special features are discussions of tools, automotive chemicals, and reference books. Finally, this book tells how to relate to the parts man and the machinist, and how to do business as a "shade-tree mechanic."

The pages of the book are enlivened by illustrations of Jenny Lander.

I hope that it will be useful and that you will enjoy it.

I would like to acknowledge the assistance of my numerous co-workers in the field of car repair. This work could not have been done without loads of help. Also I gratefully remember those soulful country singers who, along with the drink that made Milwaukee famous, pulled me through some difficult times. I would like to mention by name Darryl Schoon who gave me general assistance at a very critical time, Elaine Thayer for expert typing, and Max Knight for putting the final polish on the manuscript.

—C.J.B.

Chapter 1:

Down to Business

The miner says, "I found it!"
The psychologist says, "I feel it!"
The scientist says, "I understand it!"
The mechanic says, "It works!"

Not everybody is cut out to be a mechanic; few are, in fact. But for those who are equipped, here is a skill that will pay them well, as long as, and whenever they wish it. A bright, motivated, crazy person can make a good living, working part-time and being his or her own boss. This indeed is a civilized way to work.

I trained myself when I was on Welfare. In my spare time—on Welfare there is a lot of it—I was doing a few tune-ups for friends, working on old Fords that should have been consigned to a junkyard long ago, starting at the minimum wage. Welfare allows you to make a few bucks without penalty; a way to get you into something, I suppose. One thing led to another, and four years later, doubling my salary every other year, I now make twice as much as the average shop mechanic, with the potential to raise my wages again this year; and at the same time charging significantly less than the dealers.

Besides being lucrative, this work also keeps one in good physical shape, a fact attested to by my wife; and it is challenging, a fact attested to by myself. Needless to say, Welfare is in the past.

On the other hand, the work is dirty and difficult. If you don't like grime under your nails, which won't go away unless you substitute auto repair with dishwashing for a week, consider going into carpentry. For some reason it is rare to find a hand that can use both an open-end wrench inquisitively and a pencil literately. The literature in auto mechanics is dull—to me, anyway. The difficulty in all aspects of the repair business is at times so great that it has given me a good understanding of why a mechanic might blow his brains out. The successes, however, have been intensely satisfying.

Getting Started

As a first step into the auto repair business, you can do tune-ups using your own car for practice. Put a squib in the local paper advertising tune-ups for a particular type of car: American, VW, and so on. Quote a price around the minimum wage and the work will flow in, if you picked a popular car. If you want to work on a car like Mercedes-Benz, where you can charge higher rates than you'll ever see working on a Chevy, get an old one, tear it down, and build it up again. If it doesn't work the first time, you will either decide to go into mattress stuffing or tear it down again. Doing work over again is a basic part of learning and also develops pride in your skill. If you were already pushed beyond your limit just to get the puzzle together the first time, it's time to change. I have done some things over in an experimental fashion seven or eight times. It's amazing how fast an engine can be removed the second, third, or fourth time in a row.

Advertising is a good way to kick off or expand a business, but if it has to be sustained something is probably wrong. I started my business more slowly; by word of mouth. I even worked on the customers' cars at their own residences, initiallv. I never needed to advertise as I grew. Of course, if you aren't going to rely on honesty and careful work, advertising can take up the slack.

If you work by phone out of your own home, there has to be some way to control the noisy little bugger, otherwise it can ring you out of your mind. Many people I know use an answering service or some phone-answering device. This has its advantages. You can deal with phone business all at once, and there is no rule which says you have to answer every call.

I don't deal with the phone this way. I just don't answer it when I am under the car. You have the choice of making the business conveniently available or not, as the situation warrants. Handling business this way tends to develop a certain type of clientele: The superenergized "emergency" business is by and large eliminated; the more relaxed neighbor-dropping-by business is encouraged. Leave the quick patchups and lubrication to the service station and you'll live to an old age.

Starting from the bottom, you will see old cars with the worst problems first. The dealer mechanics look at cars with few problems on the average; in fact, often cars that may have no problems to speak of. It is obviously much harder to diagnose and solve problems on a car with a dozen malfunctions. To work on old cars is a hard school but it teaches fast.

Getting together with other mechanics has many advantages. The work of three or four mechanics can support a large shop, some of the more expensive tools, and part-time bookkeeping. Sickness of one mechanic doesn't bring the works to a halt, and mutual feedback and condolences are possible. The drawbacks are all the factors that normally make it hard for two or more humans to work together.

Specialization

There is a lot of specialization in this business. You could stop at tune-ups and make a living if you learn to do them well; and if you learn not to do them when they won't help. With a volt-ohm ammeter you can get into electrical problems. Not many mechanics are very good in this area—or in repairing carburetors. If you get good on SU carburetors, you can pay all your bills for a month in just two or three days of work and still be cutting the competition.If you specialize, probably you will have to advertise because your friends and their friends will not provide a large enough market to draw upon. For me, specialization, although potentially lucrative, is a bore. I have to be involved in new problems to keep interested and to keep from getting cynical and sloppy.

On the other hand, it is essential to have some grasp of your limitations at any particular level of development. Keep pushing—use particular jobs as an excuse to buy new tools and books, but don't take on work that is considerably beyond your present abilities. In this sense you are always a specialist.

Getting a hydraulic jack and jack stands will allow you to lift the car and work on brakes and the exhaust system. These areas, which are relatively simple but require a few expensive special tools to do the job rapidly, have been taken over to a large extent by production-line corporations that pay low wages and make money on the parts. I do brake work as a convenience for my customers even though I am unable to do it cheaper than the bigger brake-specialist outfits. For muffler work I send people to a small outfit, privately owned, which specializes in mufflers and shocks. The mechanics there do shocks as well because shocks often have to be cut off with a torch, which is the primary tool for muffler work.

Working with Suppliers

Speaking of dealing with specialists, there are several with whom you will want to develop contacts. First and foremost are the parts-store people and the machinist. Often they are housed under the same roof. My idea for a while was to find a place that would handle most of my needs for parts and machining so I could concentrate my buying power. Generally, in a year you are bringing the same amount of money to the suppliers as you are making yourself. If you are a good enough customer, a parts store might offer little extra services, such as delivery of parts, or at least pickup and delivery of large items, such as engine blocks, which are a nuisance to transport. You should get a 20% discount if you are a regular customer, whether or not you have state resale numbers or a business license. Some dealers, however, are stuffy about this.

Parts distributors are always reorganizing their lists, hence numbers on parts change. A good parts man with complete books will be able to find the correct part often through a cross-indexing search, if he is skilled and has a good relationship with you. If you learn the parts language, always have the year model and engine size in mind; and if you have the old part with you, you probably will get what you want. If you deal with a cut-rate outfit, with cut-rate help, and give them vague information, you may find that half the time you will get the wrong part.

If you find a parts man with whom you have a good rapport, stick with him. He will like to work with you if you describe your part precisely and don't get freaked out if he makes an occasional mistake. The pressure in this business is tremendous. Watch for manufacturers who use up last year's parts on this year's car; and watch for faulty parts from the factory.

If you have made arrangements through a particular parts man, continue dealing with that man asking for him by name over the phone, for example, until the arrangements are completed. When you call a dealer by telephone, the first voice will be the secretary. Ask her for "Parts." Introduce yourself as representing such and such a garage to the parts man. You will get better service. The dealer may even order a part for you without asking for advance payment.

If you wish to talk to the machinist in your parts store, ask for him by name. He won't ordinarily be answering the phone. Learn to use "machinist talk." This language consists of the names of engine parts in American English and the names of machining processes (boring, reaming, surfacing and magnafluxing, for example).

Parts stores are actually clearing houses for other businesses. There is the machine shop, which may be under the same roof or nearby, and the replacement-parts outfits, which may be the original factory supplier or a competing company. Also there are the rebuilding businesses for clutches, brake shoes, electric motors, and so on. Each parts store represents one foundry or several foundries that make automotive tools. The quality of the tools often reflects the quality of the business. Whitney's (P.O. Box 8410, Chicago, ILL. 60680) is a national parts clearing house. It's hard to predict the quality of the merchandise you will get from them. Resort to Whitney's only if your local resources can't get what you need.

Some machine shops prefer to do the complete engine overhaul themselves. You just remove the engine and put the rebuilt engine back in. It may be cheaper this way but it removes the interesting part of the work, rebuilding the engine yourself.

You can go to the smaller rebuilding businesses directly if you can find them. Look in the Yellow Pages. These smaller businesses are an invaluable resource, so develop contacts with them. You may forward business to these specialists, and when you do, tell the customer to mention who sent them. In any good-size urban area you should be able to find a specialist for any aspect of the auto. The automobile, after all, with all its supporting industries, is the world's biggest business.

Although I have stressed cooperation, a supplier's mistake should be brought to his attention; it's an unfortunate rule of human nature that if you are always the "nice guy," someone may take advantage of you. You can take for granted people closest and most valuable to you. Keep the spirit of competition alive; be always ready to go to the place that gives the best service and that has the best parts at the best price. I try to give my oldest and best customers that little extra service.

Foreign-Car Parts

Although most American-oriented parts places have started supplying foreign parts, I almost always find that the strictly foreign-parts outlets are significantly better for foreign-car parts. Also, American dealers who sell "American" cars made in other countries don't do nearly as well in keeping a good inventory for their imported vehicles.

Some foreign-car manufacturers have a virtual monopoly in the market on parts for their cars (Mercedes-Benz, Peugeot, to name some), which means that most parts, besides the simple tune-up items, are only available at the dealer at a high price.

British cars are all distributed through British Leyland, who also handle replacement parts. The British production system is decrepit, but if you know the right person you may luck out. I once got a peculiar master-cylinder kit for a Triumph from a national computerized inventory list via Jacksonville, Florida, in five days. It was the only such kit known to exist in the country. Toyota and Datsun dealership parts inventories generally seem to be well stocked, in the San Francisco area at least, as well as AMC or International, although this may be different in other areas of the country where I have no experience. VW parts are very readily available.

Some parts for VW are manufactured all over the world. The Spanish parts are terrible, the Brazilian, Australian, and Mexican parts are alright, and the German parts excellent. Generally, you get what you pay for. I got an engine-gasket set for a Volvo recently from my American parts supplier. His source for this item was the McCord company. The company makes acceptable gasket sets for American cars, but in this case the set was put together in Canada, with some English parts. It cost 60% less than the genuine Volvo gasket set. Unfortunately half the gaskets were unusable, being slightly the wrong size. This situation illustrates a difficulty particularly with Volvo—substitute parts don't seem to work well—and generally with all makes. When you don't buy parts from the dealer, you don't know whether it will be an excellent brand from England or a Johnny-come-lately outfit's product from Brazil. In a sense, most of the big brand names are clearing houses that buy products on the open market and put them into their own boxes, like meat packers buying pigs. As I said, it's a crazy business.

Most parts sources now are connected by phone or computer. This means that if a part is not on the shelf a phone call to other suppliers may get the part for you between a day and a week. Dealers usually now have computer printouts showing the inventory of all other dealers in a particular area. Even though an item may show on the list of another dealer, a quick confirming call to that parts department is in order because the list is slightly dated. Often there is a courier service connecting the various dealers. This will get you the part in a day.

As long as a particular engine or car type is in production, the suppliers who make the replacement parts will be kept in the business. When a major change occurs, with new suppliers or old suppliers with new production facilities, then parts for older models will begin to disappear. If a car company goes out of business, then parts for that type of car get harder to find. Sometimes you will find private persons selling parts from their own garage for defunct autos. If the demand is great enough, of course, somebody will start making the parts. Parts for cars like the Austin America, which was on the market for only three years, prove almost unobtainable just about the time you start to need them. If the parts can be cross-indexed to those used on other cars still in production, you are in good shape. The rule of thumb nowadays for dealers is to stock fast-selling parts for cars going back a decade. Sometimes you will run into a situation where a particular part was used in the production of the car for only six months, then the manufacturer switched to another supplier for the next five years. The dealer probably switches to the new part when repair of the old one is called for. I suggest you do the same rather than rebuild the short-lived part.

Increasingly dealers are selling cars at near cost, that is, to cover overhead and salaries, and depending on the replacement parts for the big profit. Somebody figured out that if you purchased a car part by part over the parts counter you would pay three times what you paid for the parts assembled on the showroom floor.

Specialists and Consultants

I deal with specialists who do strictly suspension and front-end work. I did not have enough repair work in this area of the car to make it worth while to acquire the special tools and knowledge to do it. In parts of the world where the roads are not as nice as in California, such work is much more important. The same is true with the automatic transmission. I guess I haven't had the interest in automatic transmissions that is necessary to explore its problems. My clientele is strange in that only 10% of the cars I work on have automatic transmissions. This is an area in which one could probably specialize quite profitably. I leave smog-inspection services to the service stations. As for chemicals, I buy kerosene and oil from the gas station, and specialized solvents, such as trichloroethylene and methylene chloride, from a chemical wholesale house. Other special greases and sealers I get at the parts store.

Finally, a mechanic needs a consultant in his own area. You will be a pain in the neck to people who are depending on you, if you are unable to say "I don't know," when you don't. If I run into a difficulty which I can't solve, or if there is some question whether I have been responsible for shoddy work, I take the car to a mechanics garage run by a fellow who has been in the business for several decades. He charges more than I do, but at some point solving the problem is more important than cost. Besides, when you get to the stage where you need consultation, you expect to pay more. The only question is whether the consultant does in fact know more than you do. In this game, years of experience weigh very heavily. Two of the people whom I consult are the second generation of the family in the business.

Generally speaking, the best businesses I have found have been those run by one man or perhaps a father and son. They have been businesses in which up to fifteen people are employed, but are still run by one man or woman. He or she may not work on the counter, but is active somewhere nearby. If you become a regular customer, you will undoubtedly get to know the boss. In so many areas of this business things are so diffuse, companies so large, that the buck gets passed on indefinitely. It is nice to deal with a business where you can find a person who will stand up and say, "I am responsible for this, what is the problem?" I am not talking about assigning guilt, but simply about wanting to find the solution to a problem.

Before I get caught making generalizations about the virtues of small businesses and the vices of dealers, let me say the obvious—not all small businesses are reliable or all dealers sloppy and expensive. All it may take for a business to go to pot is an expansion that backfires or an owner who gets a divorce.

In general I find that mechanics who really know how to rebuild generators and transmissions can work their own hours, charge top dollar, and still have a constant two-week backlog of work. And I've never seen the best places flooding the market with advertizing either. In the mechanics business, skill is always rewarded well as long as you don't work for the dealer.

What to Charge

This leads me to the area of determining what you yourself are worth. When you are self-employed in any area you have to give thought to this. You can start with the minimum wage. The most inexperienced laborer should at least be paid the minimum wage. The dealer, I think, establishes the criterion for quality of work and for the hourly wage. He usually pays the mechanic about a third of what he is charging the customer for hourly labor. Even with all those super radar scopes (which I hardly ever see being used, by the way), the quality of most dealers' work is only average.

The reasons for this are simple. The dealer mechanic doesn't have to be directly responsible to the customer for his work. The dealer mechanic is underpaid if compared with, say, a bus driver. And the dealer mechanic generally is a parts replacer. He doesn't rebuild units (the dealer makes more money selling new parts) and therefore doesn't really know how things work as well as somebody who rebuilds malfunctioning units. I don't know alternators as thoroughly as I might, because I often have them rebuilt for me.

Many mechanics garages charge a labor rate that is close to that charged by the dealer. They beat the dealer by doing better work and by offering rebuilding services. This lowers the total bill.

In both situations overhead and insurance have to be paid for. If you can cut out the overhead by working out of your own backyard, you need only work roughly half as long for the same profit, or you can make roughly twice as much. I would concentrate on the former advantage, because it would allow you to work less in the winter, for example. I find I can't work longer than around three hours at a stretch in the winter before I've lost too much body heat. Of course one can heat the work area to a comfortable temperature; but any way you cut it, it's going to cost more to work in the cold. I also use the extra time to be with my son and to do a lot of the cooking. Two nice balances to mechanicking.

Actually the hourly wage is only a rule of thumb for me. I also have a standard fee for common jobs like tune-ups. I generally make more than my hourly wage on these jobs because I am very skilled at them, and so do them rapidly. The set fee is still planned to be competitive with the rate charged by local garages. This salary advantage is necessary to counterbalance the more interesting jobs for which I might not feel I can charge my regular hourly rate and still be competitive. One way to get interesting work, or to expand your areas of competence, is to underbid the market average price for a particular job. On a big job, the second time around on the same work you can probably double your profit for the same price charged. If you develop a skill in an area where there is essentially no competition, you can charge whatever the market will bear.

One has to be careful about raising prices. Not too slowly, not too rapidly. You have to keep up with inflation at least, but if the increase is too great, meaning you waited too long, you will lose too many old customers before you find new ones willing to pay the higher prices. There is an infinite number of levels on which to do business. Some people wouldn't pay more than the minimum wage per hour labor, whereas others would be overjoyed to find somebody to do competent work at union rates.

I have found that as I raised my prices I got better and newer automobiles to work on. My skills without a doubt were increasing but, besides, many people wouldn't trust a fancy car with anybody who didn't charge enough. Stranger in a strange land.

Actually most places, especially the dealer, make a profit on the parts in addition to the labor. There are three or four wholesale prices and the so-called retail price for parts. The customer pays the retail price, and all others who handle the part, unless they are working for the man on salary, get a percentage. You can decide whether you want to make 20% or so on parts. I usually don't add the 20% unless it took some special effort to get the parts.

It is the final bill in which the customer will be interested even before the work is done. This means the mechanic should be able either to make an estimate or give a quotation. Most mechanics give a quotation, which is a figure that is supposed to be held to. Usually the quotation reflects the *most* it would cost a skilled practitioner to do the job. Except for my set jobs, I give *estimates* which may vary by as much as 25% more or less. If your final bill is lower than the estimate nobody will complain, but if it is higher it is somewhat unfair to the customer. With old customers I don't mind taking deferred payments. If you aren't a good mechanic I wouldn't recommend it.

A final note about what prices to charge. My rule of thumb for charging prices as well as for the level of business to accept, is to try to operate normally at around 80% of full capacity. This means that I charge 80% of what I think I could get for that particular job. The reason for this is to allow room for temporary expansion. It's nice to know you can work more or charge more in an emergency. It can turn your brain to oatmeal to be working flat out and have somebody return with a blown engine you just rebuilt.

Responsibility

This leads to the question of responsibility on work that comes back. It is this thorny problem that keeps many a shop mechanic making $8 per hour under the "protection" of the dealer. The fact of the matter is that if you wish big brother to protect you from your mistakes it will cost you much more in lost wages than if you plan it for yourself. If you wished actually to make $10 per hour, you can charge $12.50 per hour, the difference being your insurance policy. Since the average job is only a day, the most you can usually lose is a day's labor. This is a rather small loss if you are making your basic monthly bills in a week.

When big corporations deal with big corporations or when individuals deal with corporations, the lawsuit is the common mode used to resolve differences. I use the technique of haggle and compromise with dissatisfied customers. No matter how good you are, work will come back. And when it does, there is the question of who is responsible, who is going to pay. I take each case as it comes. Since I work with this problem in mind, following rather compulsive procedures, I keep returns to a minimum. But there are all those remanufactured parts.

Some returns are due to new problems which only appeared when the previous difficulties were solved. When this is discovered, there is no problem, at least for the mechanic, just more work. Sometimes you have left a screw loose or something similar. No problem here either. You just fix it—no charge. But what about a situation where you did everything you were supposed to do, and the remanufactured part fails? You probably can get that part replaced by your parts distributor; but then what about all that labor? I will generally argue for settling 50-50 with the customer. It isn't the customer's fault, it isn't yours, and it isn't the parts distributor's fault. It's somebody two or three steps removed from the parts distributor. When you shove the bad part back to the parts distributor for a good one, you hope he will shove it back down the tube to the source. If the customer and the mechanic share the rest of the loss, then nobody bears too much of it. I have never lost a customer using this approach. On the other hand, I don't feel responsible for damage created by the owner's abuse of the vehicle; and ignorance is no excuse.

The consumer movement has ended the time when godlike authorities could issue absolute pronouncements about the causes and cures of ailments. Whether you are a doctor or a mechanic you should learn the probabilities of success for the things you are fixing. The consumer won't like to hear the truth: that there are no failsafe solutions. But if you give the public a sugar-coated pill and it doesn't work you are liable to have a lawyer ringing your bell. The fact is that even new parts may fail. Reducing the probabilities of failure usually requires more money and often more time. A practiced mechanic will know how to play the odds better than a greenhorn. Try to inform the customer ahead of time what the odds are for failure with any repair procedure, to the best of your knowledge.

The Public

Whenever you are self-employed you have to deal with the "crazy" public. Crazy in this sense means ever-changeable. You get the bargain hunters who carefully examine all facets of the solution from one hundred miles to an inch. Then there are the establishment people who are very nervous dealing with a specialist who doesn't wear a white coat and who talks about probabilities rather than certainties. The anti-establishment people for whom the shade-tree mechanic is the only acceptable solution to car problems are, of course, a mainstay of the independent. Make a mistake or charge what some of these people think is too much, however, and you are liable to have a moral crusade on your hands. Then there is the rather large subgroup of women looking for a mechanic who won't put the make on them or treat them like dumb dumbs. Or how could we forget the male car owners who would have done the job themselves if only they had the time, and tell you exactly what to do. In regard to the last situation: always make your own diagnosis. A type of customer I consider a ripoff is the person who does a "tune-up" in his backyard, then comes to you to "just finish it off" for him. "It will only take a few minutes," says he. It never does. Charge the full rate says I. He can watch and learn.

I have always allowed myself to "fire" a customer. I am not running an outpatient clinic for paranoids. A paranoid in this business is a bull in the china shop. There are clip joints run by paranoids, which are perfectly suited for paranoid customers. Let them join and do battle with one another, and stay aside if possible. Generally I find people to be easy to get along with, and the problems in customer relations can be solved by an open discussion of the obstacles that have to be overcome. Most people have only a vague idea of what's involved in car repairing.

Other Business Matters

Other nice things about this business are that you can deal entirely in cash. Also, you might lose customers for one reason or another, but you never get "fired" by your own business.

Two bookkeeping hints. If possible don't handle any piece of paper more than once. Also, if you pay your taxes in January instead of April, you don't have to go through all that quarterly withholding-tax bookwork.

If you work in a residential neighborhood, you may be violating zoning regulations. These regulations are not usually enforced unless the neighborhood insists. This means you will exist at the sufferance of the people living nearby. Keep your work area clean—don't let sick cars fill up the streets. Give special attention to neighbors' cars. Neighbors can become your best customers. In addition, crime is often reduced if there is somebody "physical" like a mechanic around all day to keep an eye on things. You can become a major asset to a neighborhood.

The Future

There has been a resurgence of the shade-tree mechanic in the past decade. The increasing costs of repair at the dealer on the one hand and the rather low wages paid to dealer mechanics on the other have encouraged the public and the mechanic to meet more directly. In addition, the gas companies who are now looking for increased profits at the gas pump are forcing the gas station out of the repair business so that they can concentrate on a greatly increased volume of gas pumping. But the biggest area of change will be the basic design of the automobile itself.

There are several forces operating in the car market these days. With the cost of gas going up faster than our salaries, a more fuel-efficient engine is making its appearance. Even though at this writing some of the American push-rod 6- and 8-cylinder-engined cars are still a good buy when all costs are considered—especially on the used market—their days are numbered.

As a concept, the push-rod engine was developed in the thirties, but it really dominated the market in the years following the Second World War. It was perfect when fuel was cheaper and the market still expanding. The first owner would use it for two to five years, then it would pass to the second-hand market where it would be of use to others for another five to eight years or more. This became a novel method for spreading the wealth around. The second, third, or fourth owner would get the essential use of the auto at a much reduced cost, although without receiving the magical benefits of newness that inspired the first owner. The idea of obsolescence through styling changes pioneered by Chevy helped this process enormously.

The engine is now being reduced in size and changed in basic design from the 6- and 8-cylinder push-rod types to the 4- and 6-cylinder overhead-cam engine. The change from the push-rod to overhead cam allows a reduced size without reduced power output. For example, the older VW push-rod pancake engine got around 50 horsepower out of an engine with the same volumetric capacity as the new overhead-cam engine. The new engine gets 70 horsepower and as a result has better gas mileage and performance. It simply uses the same volume of fuel more efficiently. Roughly, this same increment of improved efficiency was gained when the old flathead engine gave way to the push-rod engine in the early fifties.

What is lost in this evolution is durability. The newer engine achieves the higher output at a higher r.p.m. (revolutions per minute) range, which means the engine turns more frequently during the course of a mile than the older designs. Obviously the engine will wear out sooner. The cost of repairs has gone up because the newer engines are more difficult to work on, although part of the problem here is that there is a shortage of mechanics skilled in these new designs.

One way to avoid the new technology is to reduce the weight of the car, which is done by switching from the use of steel to aluminum and plastic. Cars are also becoming smaller and more boxy to make more efficient use of space. A lot of weight is saved by having a front-wheel-drive design. The transmission and differential occupy the same housing, eliminating the driveshaft completely. More space is gained by this change as well.

Besides weight, a great deal of experimentation has been done and will be done on the engine's carburetion. This work introduces another of the forces causing change in the automobile, the need to control harmful emissions. Unpleasant emissions marked the end of another earlier mode of transportation—the horse. At any rate, more efficient carburetors will increase gas mileage and reduce smog. The more conservative American manufacturers have merely tinkered with the old standby, the carburetor, making it a plumber's nightmare.

European manufacturers went to electronic fuel injection as early as 1968. This system was an electrician's nightmare. However, a new type of fuel injection has been designed, which has the promise of a good future. It is not electronically controlled but relies on simpler mechanical control devices (not like the mechanical injection system on the diesel Mercedes). Being so radically new, mechanics will have difficulty with it for a while, particularly those who won't or can't read.

Another force in the car market is that it has reached or shortly will reach saturation. This means that sales will hinge on replacement of the existing stock of cars now on the road (roughly 100 million) at a rate of around 10% a year. In a growing market some inefficiency in a company can be tolerated, but when a plateau is reached the marginal companies will go under. One way to avoid this temporarily is to make cars that don't last as long. As I mentioned above, the newer engine designs will have a shorter life.

If we can infer the motivation of a company from the design of the engine it manufactures, then the 30,000-mile car with a completely sealed engine—you only put in gas—reveals that motivation: built-in obsolescence. The Chevy Vega comes to mind. It has an aluminum block which is not intended to be rebuilt—only replaced. The alternator mounting is incorporated with the thermostat housing, making it almost certain that a water leak will develop in half the normal time. The timing belt is tensioned by the water pump and it runs the fan. The inconvenience and costliness of the repairs—and the repairs of the repairs—resulting from these "clever" innovations is too depressing to contemplate. There is a weakness in the cooling system, a cardinal sin with an engine that has an aluminum block. These are just a few examples in a car that seems to me to represent the concept of the nonrepairable automobile.

The Vega is not the only example of such car design. Toyota—the "tough one"—seems to be made of some sort of self-destruct pot metal. But don't complain, folks. There is only tradition on your side, the tradition that a car should go ten years rather than five. The ten-year cars are still around, they just cost more. Of course the first owner of a cheap car doesn't feel the first crunch. It will be the later owner who will suffer first. However, with a little experience in these cheap cars their resale value will drop and this will hurt the first owner.

Not all contemporary engine designs are conveying the same message. VW for years revived one of the basic principles of the model-T Ford. A simple economical car, which was changed not to make it obsolescent but to improve the original design. In my opinion, VW pushed the "beetle" concept about five years too far with the "super beetle" monstrosity, but the company redeemed itself with the Rabbit; at least to the extent that intention can be inferred from engine design. The company's philosophy of advanced technology is its competitive edge in the market place.

The Rabbit is thoroughly modern in design: four cylinders, overhead cam, front-wheel drive, and fuel injection. So far, these features don't distinguish it from half a dozen other European cars on the market, which are exceedingly difficult, therefore expensive, to work on. But the following features reveal the manufacturer's intent: The oil pan can be removed with the engine in the car, allowing the repair man to check the main bearings and do a ring job rather easily. The cam directly activates the cam followers, which in this case are cams placed over the valve springs. This means that the valves will probably not need adjusting before a valve job is required. The mechanic can remove the transaxle without taking out the engine and does not need a special tool to remove the clutch. Few front-wheel-drive cars have this feature. The difficult electronic-brain fuel-injection system has been replaced by a much simpler constant injection system that is no more complex than the old carburetor B.S. (before smog). The newness of the design will make it difficult at first for many mechanics, but even this disadvantage is eased by the fact that the finest service manual I have ever read on any car is available for the Rabbit and Sirocco, published by Bentley.

Although some terrible deficiency may develop in the Rabbit (the carburated Rabbit has had its problems), the intent is clear, just as the intent of the manufacturer is clear in the engine design of the Vega. The Vega costs less than the Rabbit; in the showroom at least.

Most new-car reviews aren't worth the paper they are written on. For one thing, major engineering flaws don't usually begin to appear until two or three years after the car is on the market. The test track reports of cornering, acceleration, and braking are irrelevant to the ordinary driver. If people drove the way those test drivers do, our lives would be in mortal danger. The main thing we would like to know is durability, but to judge this takes time and even Detroit doesn't have enough of that. When a new model comes on the market you can bet it is as green as the wood you get at the lumber yard these days.

It is sad to see how incapable Detroit seems to be in coming up with new technology. Electronic ignition has been the only real contribution to the mass car market from the Big Three in the recent past (I'm ignoring smog devices because of their questionable value in some cases and, more basically, because the manufacturers were forced into these changes from outside the industry). One of the reasons for the failure to lead in technology is that the big American companies that sell cars are producers and merchandizers—not innovators. The basic innovations in the four-cycle internal-combustion-motivated conveyance occurred before General Motors et al., brought about by individual genius or by paired genius. The big guys bought up the best of what was available for a mass-consumed auto in the 1930's and the individuals withered away. Even today, most innovations in American cars of the Big Three come from purchases in Europe or Japan. Often they buy out the little company that created the product. Detroit cannot be lauded for introducing basic innovations but it can be praised for putting together a very complex machine, which operates quite reliably even in the hands of a mechanical moron.

The difficulties of improving on the automobile are enormous. Enzo Ferrari, who designed more World Class Champion cars than any other person alive or dead, felt we have run out the string on the internal-combustion engine. When his refinements are fully translated into the mass-produced car in the next decade where do we go from there?

Is the End in Sight?

Perhaps the Big Three are so resistant to change because they feel the days of mass transportation by automobile are numbered and therefore the problems are inconsequential. Many cars on which I have worked recently suggest that the manufacturer is planning to be in another business before long. To get things in better perspective let me do a quick sketch of the automobile culture as I see it.

The railroad, which seems to be the most energy-efficient form of transportation, reached its peak of growth in the late 1920's at the same time that thousands of private and public decisions were being made to build an automobile culture. Most everything made since then, which includes most everything we have, has been made for, with, and , in rare instances, in spite of the automobile culture. Our food, housing, and work are planned around the use of the infernal contraption. In fact, our riches whatever they may be, stem from this incredible achievement of the Industrial Revolution.

Oil embargoes, smog, inflation, and Ralph Nader notwithstanding, cars are safer, jazzier, and easier to own than ever before. Just about everybody can still go into the market place with a couple of hundred dollars and pick up a ten-year-old wreck with a few thousand miles left in it, get in, turn the key, and have a live infernal combustion engine at his or her foot.

I am reminded of the Indian sage who points out to his disciple that the earth is covered with leather if he always wears shoes. I wonder what he would say about having an engine at the end of one's foot. Keep it in tune, perhaps.

I came across a world map published by the Freedom Foundation. The countries were in white, grey, or black, to represent those which were "free," "partially free," and "enslaved." It's a rather mind-boggling idea, but I noticed a high correlation between the "free" countries and those with a high level of automobile production and use. The "free" area epitomizes what we call Western Culture, with Japan thrown in. Do you suppose this means that a change in politics would change our automobile dependency, or does it work the other way round? If we could introduce a cheap Chevy into mainland China, would they become capitalists?

I don't know the answer, but I know that we are dealing with an addiction which we soon will no longer be able to afford. Being a mechanic, I hear it all the time from all walks and economic levels: "I'm dependent on my car." The 1920's vision of freedom is being reduced to an absurdity. The "automobile freedom" is becoming an Albatross. The greatest hidden cost of wealth is the need to establish limitations. When you are poor the limitations are established for you.

As I see it, the external problem is that anything produced in the quantity of the automobile stresses the reserves of basic raw materials and creates undesirable by-products. All natural processes unguided by an overriding principle will expand to the limit of their raw materials and poison themselves with waste. The limit will be reached abruptly or gradually depending on the perspective of the observer.

Complicating the external problem is the psychological one of addiction. Detroit is the dealer, so we can't expect the dealer to change our habits. The best he can do is cut the purity of the drug so that we can continue to "afford" it. The government being a beneficiary of automobile success cannot be counted on for real assistance on this problem. But the habit can be kicked by individuals just the way any other habit can be kicked. Recognizing it as a habit is the first step—a deeply rooted emotional habit with its own unique rhythm. Can you imagine how rich we could be if we didn't have a car to support? We could afford, for example, to support a neighborhood grocery within walking distance instead of needing the supermarket.

Unless we go mad, and like the lemming drive right off the edge of the energy cliff in our Merry Oldsmobiles, with only perhaps a last moment of regret that pleasure often seems to be followed by pain, there's hope; there is always hope. The infernal combustion engine could serve a lengthy continued usefulness in a more subordinate role rather than as the "hot rod" people mover. Electronics and communication could do more to bring the world to us, so we don't have to drive around to find it. And as a primary economic and emotional theme, electronics could certainly replace the last gasp of the Industrial Revolution.

It's a new day. New movers and shakers are coming to the front. Nevertheless, whatever happens, short of total collapse of our culture, competent and honest general mechanics will be needed for some time to come—and hard to find.

Chapter 2:

The Tools of the Trade

There once was a Ford made in Cheswick,
had holes in its pistons, and valves stick.
We opened the hood
and did what we could
with tongs, and a spoon, and an icepick.

—Roger Prichard (b. 1940)

Hand Tools

One of the nicest things about being a mechanic is that you have a good reason to buy lots of tools. Proto, Snap-on, Bonney, and a few others make fine tools for auto mechanics, but when compared with Sears Craftsman tools they are overpriced. Sears's top line, which are always labeled Craftsman, are very good quality and backed by a lifetime guarantee. If you break one of their tools you simply walk into one of their stores and exchange it for a new one. For the working mechanic a lifetime guarantee doesn't mean much if he has to chase all over hell or wait six months to get what he needs.

The flea market offers a limited opportunity to buy tools, but about the only true "find" is when someone is cleaning out the tools of a sadly departed brother mechanic. Skilled craftsmen hardly ever give up their basic hand tools unless they are getting better ones, and it is advisable to get the better ones in the first place. Older-vintage carpentry tools are often superior to what one can get at the neighborhood hardware store, but I have not found this to be true of automotive tools. The newer the better. Don't buy used hydraulic jacks or marginal lifting equipment of any sort. One's arm or leg, or even life, may be in the balance. If you are buying a chain, check to see if the links are stretched. If they are, don't buy.

Don't buy those 200-tool discount-house specials; discounts are usually found on inferior tools. You will end up paying more in the long run in skinned knuckles, tool replacements, and botched jobs. Even though some money can be saved on Sears tool-set specials, I advise against these as well. They include tools you may never use. Using tools is a very personal thing. There are tools I never use, which are frequently used by other mechanics and vice versa. Make your own selection.

My recommendation is to start with the top box of a fancy Sears tool-box set (either a 6- or 8-drawer box). Later you can add the rolling cabinet and drawers in between the two. The tool box protects the tools from the weather, discourages thieves and, perhaps most important, imparts some order to things. When you have thirty tools this is not a problem, but when you have three hundred it is essential to evolve your own quick recall system. One can easily spend as much time finding lost tools as doing the job itself, which is no way to run a business or a life. To keep good order is not so onerous if the system is your own creative invention. I put my own wooden dividers into the drawers.

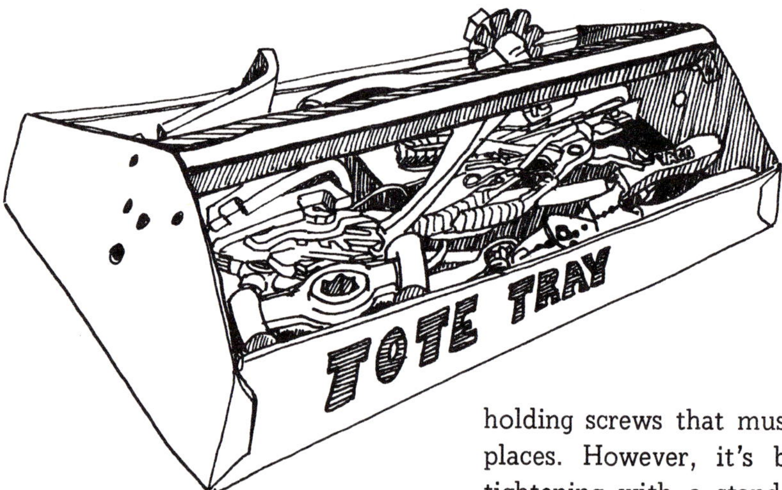

The described tool box has a tote tray that fits in the top. It carries a limited selection of tools to the job. The tools that stay in my tote tray could be called my basic tools. A description of these tools follows.

Screwdrivers

I have at least three dozen screwdrivers. Everything from jeweler's to Sears's biggest, from offset to impact. One shouldn't use them for prying or chiseling, but sometimes one does. The blade size should be the same width as the screw slot. The end should be squared, not sharp. I recommend screw drivers with square shafts. Such shafts enable you to use a crescent wrench, press down on the handle, and turn with the wrench. This usage provides a great increase in torque. A wedge-type screwdriver is useful for holding screws that must be started in difficult places. However, it's best to give the final tightening with a standard type. Sears drivers have two limitations. The blades aren't honed, hence they don't hold the screw slot well, and the handles become sticky with brake fluid and with some solvents. Proto drivers do not have these drawbacks.

Additional Screwdrivers

Impact-Driver Set.—This tool is used with a hammer for breaking loose tight screws. A hammer blow can deliver through the impact handle a

Screwdrivers in the Tote Tray

Blade width	*Shaft length*	*Size*	*Type*
1/8"	4"	Extra Small	Standard
3/16"	4"	Small	Standard
1/4"	4"	Medium	Standard
1/4"	1½"	Stubby	Standard
No. 1	3"	———	Phillips
No. 2	4"	———	Phillips
1/8"	3"	———	Wedge Type

momentary high-torque blow, which far exceeds the twist one can apply on a regular screwdriver handle. This tool is essential for the countersunk Phillips screws used on car-door hinges and for aluminum screws, often used on motorcycles. An impact driver is also useful on the metal shrouding covering a VW engine. One can tighten screws to a far greater extent with this tool than with a regular screwdriver. The impact driver usually comes with two sizes of standard screw blades and two sizes of Phillips. You can also use sockets with this tool.

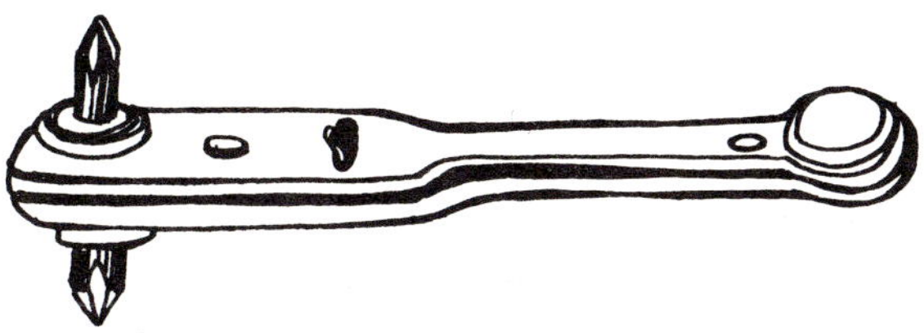

Offset Screwdriver.—This type is used in tight corners without clearance for a screwdriver handle. There are types that ratchet, which makes the work easier. It is easy to destroy a screw slot with these. One way to avoid this is to keep the blade in the slot and on the same axis as the screw.

Wrenches

The open-end wrench is the most universal tool for hex-headed nuts and bolts. It is almost the symbol for auto mechanics. Its limitations are that the jaws tend to spread when great pressure is applied, and that it applies pressure on only two flats of the nut. A tool made with inferior steel will spread quickly and stay spread. You can choose combination wrenches that have a box on one end and an open end of the same size on the other. I use separate open-end and box wrenches, both of which have two sizes.

Where two open-end wrenches are required to do a job, such as on the nut and bolt holding the starter motor in place, the nut size and head size are often different. Sometimes the two sizes are quite different as on some gas-line-to-carburetor fittings. In other cases, as in the generator or starter motor, an odd-size nut will be used. This is where the 19/32" x 11/16" combination open-end wrench is usually employed.

There are other cases where two nuts of the same size have to be manipulated simultaneously, one being held as the other is turned. In this situation a 6" crescent wrench is handy. It opens to all the normal 3/8" drive socket sizes. Its weakness is that it tends to change size as you use it. Therefore it's a good idea to keep retightening it to avoid skinned knuckles and rounded nuts.

Additional Wrenches

Box-End Wrench.—The box-end wrench, which has a closed ring, doesn't suffer from the problem of the spreading jaw, but it cannot be used in as many places as the open-end wrench. You can choose either 6- or 12-point box wrenches.

Wrenches in the Tote Tray

Open-end standard	*Open-end metric*
3/8" x 7/16"	6 x 8 cm
1/2" x 9/16"	7 x 9 cm
19/32" x 11/16"	10 x 11 cm
5/8" x 3/4"	12 x 14 cm
	13 x 15 cm
6" crescent wrench	17 x 19 cm

The 6-point wrenches apply greater pressure on the nut or screw head, whereas the 12-point wrench can be put on the nut in twice as many positions. A lateral clearance of only 15 degrees is necessary before the 12-point box-end wrench can be repositioned on the nut. Both long-handled and short-handled box-end wrenches are useful. The smaller sizes (7/16" x 3/8") are good for brake-bleeding nipples. There is a special one made for VWs, which I recommend. If you use anything else you risk breaking off the nipple, which is expensive and time consuming to repair.

Offset Wrench.—This is an open-end wrench that opens at 90 degrees to the handle and in line with the handle. It is usually purchased for jobs where space is tight, for instance on the inside carburetor-to-intake manifold nut on a conventional VW. The nut is most easily removed with a small 13 mm 90-degree offset open-end wrench. It may be difficult to find this tool. There is a curved 9/16" x 5/8" manifold wrench sold by Sears that I have not found to be particularly useful. In the only situation where my regular wrenches wouldn't work (on a 1950 Studebaker), this wrench wouldn't either.

Tappet Wrench.—This is a long, thin open-ended wrench used in tight spaces, for instance when the tappets of a flat-head engine are adjusted, where one wrench must hold a nut while the other turns. Being thin, they engage only one nut at a time. They are valuable at times.

Ignition Wrenches.—They come as a special set of offset open-end wrenches in small sizes such as you require in distributor work. Initially you can probably substitute a small crescent wrench.

Flare-Nut Wrench.—This wrench is indispensible for loosening brake-line or gas-line flare nuts. It is a cross between an open-end and a box wrench. Flare nuts, being brass, round off rather easily if they are a little tight, but I've never had any problems when a flare-nut wrench was used. Sizes 3/8" through 9/16" ought to cover most flare nuts used in automobile construction.

Crescent Wrenches.—They come in handle sizes from 4" to 16". I use a 6" handle, which has a jaw capacity up to 1 5/16". These wrenches are used mainly to supplement the fixed wrenches. Their limitation was mentioned above.

Crowsfeet.—These are open-end wrench heads that fit on the end of a 3/8" drive extension and then to a ratchet handle, and are useful for getting at nuts that are recessed or under a ledge such as the hold-down clamp nuts on many distributors. There are other special wrenches made to do this as well. If you have crowsfeet, 1/2" and 9/16" sizes will probably take care of most uses of this tool.

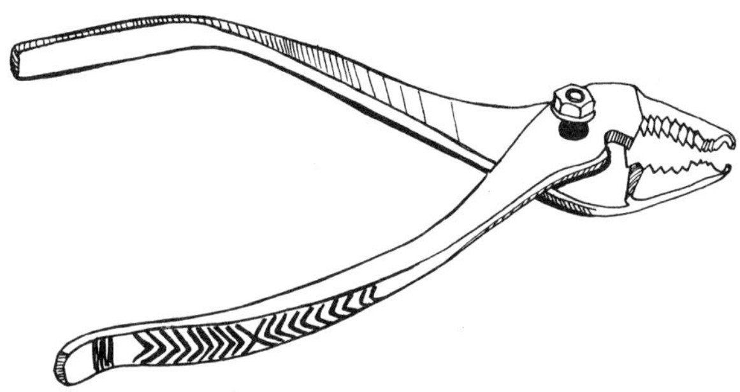

Pliers

Pliers.—Just as the open-end wrench may suggest the auto mechanic, a pair of slip-joint pliers suggests to me the kitchen-drawer and the weekend handyman. Pliers are good for handling springs, pulling dust covers, and a few other purposes. I grind down a standard pair of slip-joint pliers into a parrot's beak.

Pliers in the Tote Tray

One pair standard slip-joint curved-jaw pliers, which have been ground down to a parrot's beak

One pair straight-jaw battery pliers

Additional Pliers

Vise Grips.—These are locking pliers, which give you a third hand. Get a couple of these early in the game. I have two small ones and one big one. These can be clamped on a clutch or handbrake

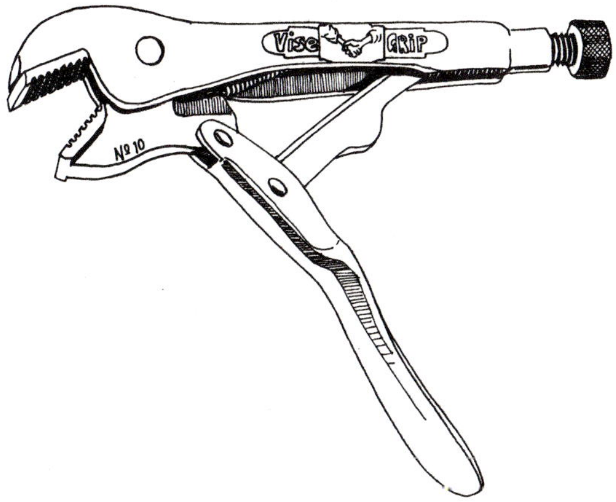

cable to keep it from turning while you turn the adjustment screw. They can be used in a pair, one clamped on each side of a bearing so that two screwdrivers can be used to pry out the bearing. Also, they can be used to keep the head of a bolt from turning on one side of the firewall while you loosen the nut from the other side, or to clamp onto the stub of a broken stud to turn it out of the block. It is the tool most associated with difficult and unconventional work.

Hemostat.—This is a mini vise grip, good for holding small things in place. On slip-on coil connections, such as are used on new Ford products, I clip a hemostat on the post, then put the connector back on the remaining stub. Then I hook my tach-dwell meter (see p. 20) to the hemostat making sure it doesn't ground to the engine. This tool, used commonly in surgery, may be found in electronic supply stores.

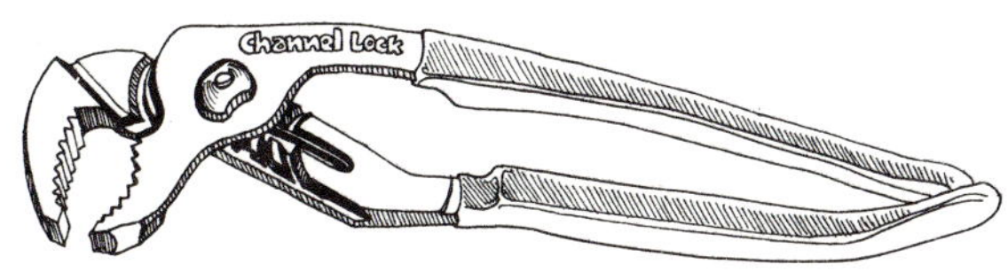

Channel Lock Pliers.—This is the tool for pulling the dust cover off the front axle. Its angled teeth permit its use also somewhat in the manner of a pipe wrench.

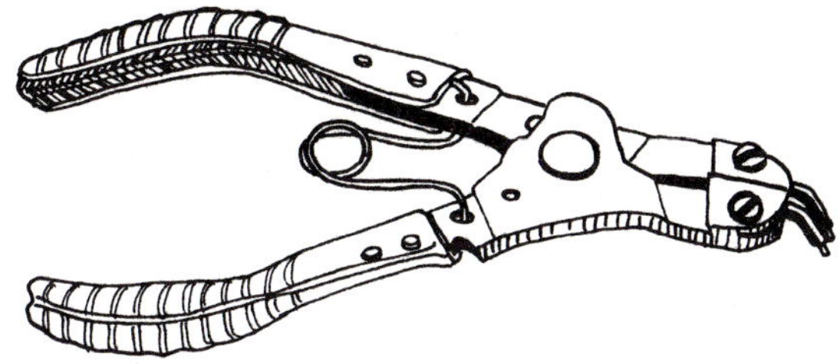

Snap-Ring Pliers.—These pliers are used to remove internal and external snap rings. These rings are typically found holding bearings in housings or on shafts. K-D makes a set with replaceable points. Another type of retaining ring is an outside ring without holes in it for the snap-ring pliers. This type is removed with lock-ring pliers.

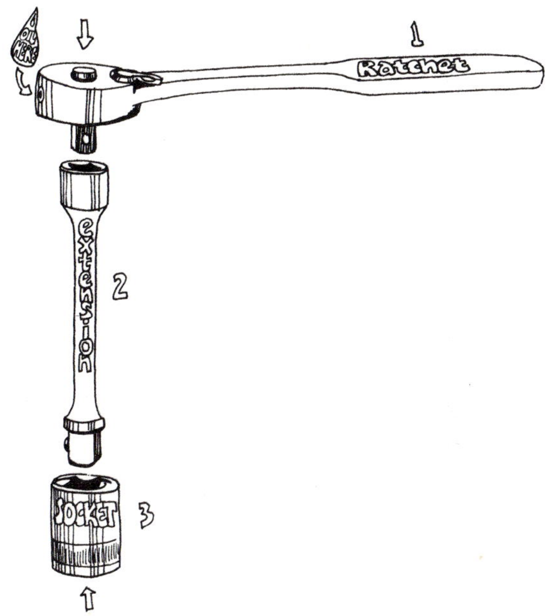

Ratchets.—The bread-and-butter tool of the mechanic is the ratchet handle and its accessories. The ratchet handle allows the user to tighten or loosen a fastener without removing the wrench. This gives greater speed over the conventional open-end or box-end wrench and so it should be used wherever possible. Although I do use a lot of Sears tools, I have a Snap-On ratchet handle because of the beauty of its design, its good feel, and its reversing lever, which points to "on" and "off." Things can sometimes get so dizzy that it's hard to remember clockwise and counterclockwise.

The sockets that go with the ratchet handle are similar in application to other wrenches, but can reach into recesses that those wrenches cannot. Sockets also are available in 6-point or 12-point. My preference is for 6-point because it gives a firmer contact with the nut or bolt. Since one runs into some strange nuts in this business, I have a full set of American standard sockets as well as metric and Whitworth sockets. You want to use the best-fitting socket no matter what the nut size is supposed to be. A 9/16" American standard bolt may be reduced in size because of rust. A 14-mm metric socket is slightly tighter than a 9/16" American standard socket, and a 1/4" Whitworth socket is tighter yet. The next-smaller size is the 1/2" American standard socket. On socket sizes over 1" I find American standard sizes will cover all applications. Generally one is working on either an American standard or a metric car, so I carry only the appropriate wrenches and sockets in the tote tray. Sockets can also be used as drift punches to install grease seals or wheel-bearing races.

Ratchet and Accessories in the Tote Tray

- 3/8"-drive ratchet handle
- 3/8"-drive normal socket set (metric or American standard 6-point)
- 3", 6", 12" extensions
- universal joint
- spark-plug socket

Additional Ratchets and Accessories

Sockets.—A 1/4"-drive set is nice, preferably in its own little box to keep all the little parts together. It is also useful to get screwdriver, Phillips, and Allen-head or hex-head sockets with this set as well as extensions and a universal joint. I also use 1/2"-drive sockets from 9/16" up to 1" with a 1/2"-drive ratchet handle, extensions, and universal joint. These will be necessary when you progress beyond tuneups to brake jobs, valve jobs, and so on. Instead of your getting a 1/2"-drive breaker bar or flex handle, I suggest the 3/4"-drive handle and an adaptor to convert the 3/4" drive to a 1/2" drive. Proto makes such a socket converter. This allows you to get the few giant sockets you may need in 3/4" drive but to use the same handle for the smaller sockets. With this heavier breaker bar you can add a cheater pipe. I use a 4' pipe on my 3/4" bar and easily get the 200-250 lbs. of torque necessary to remove the VW flywheel gland nut. From an inch on up I use 3/4"-drive sockets. These are useful in engine overhaul. They are expensive, so you might buy these one at a time as needed. I might note that cheater pipes or extension handles are considered bad practice because the length of the ratchet arm or wrench is supposed to give the leverage appropriate for the size. I agree with this most of the time, but the guys who wrote those rules weren't stuck under a dirty car somewhere with limited access and awful leverage.

Deep sockets are for long rundowns before the nut is tight, for example on U-clamps. Some deep sockets are not grooved all the way down the barrel. I would avoid these.

If you are working on a Mercedes-Benz, and to a lesser extent on a Volvo or VW, a set of Allen-head or hex-head sockets in metric sizes 9 mm to 17 mm is necessary.

Gauges

The feeler gauge and the wire gauge are the primary measuring instruments in auto mechanics, and they are in my tote tray. The feeler gauge

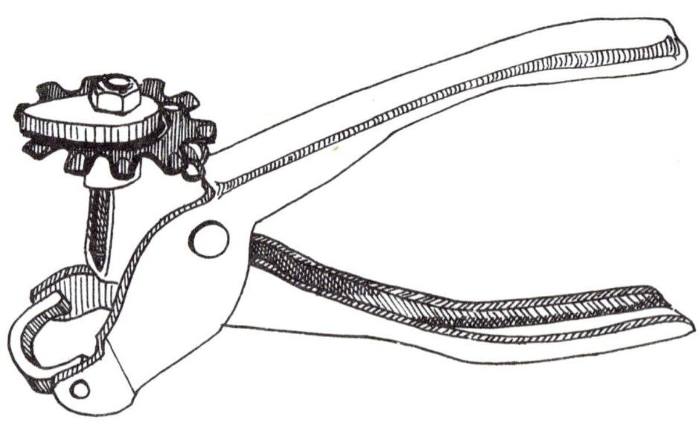

measures the space between two flat surfaces, such as the valve lash, which is the space between the rocker arm and the end of the valve. A feeler-gauge set is a stack of thin metal blades that are progressively thicker by .001" increments. I use a go-no-go gauge. The tip of this feeler is thinner than the rest of the blade by .002". Therefore if you are using the .008" blade, the the tip is .008", and the rest of the blade is .010". When the tip of the blade can move between the adjustment screw and the valve end, but cannot move over the .002" shoulder to the thicker portion of the blade, then you are within .001" of the correct setting. It is hard to find this type of feeler gauge. You may have to order it. In my experience, mechanics tend to set valves too loose.

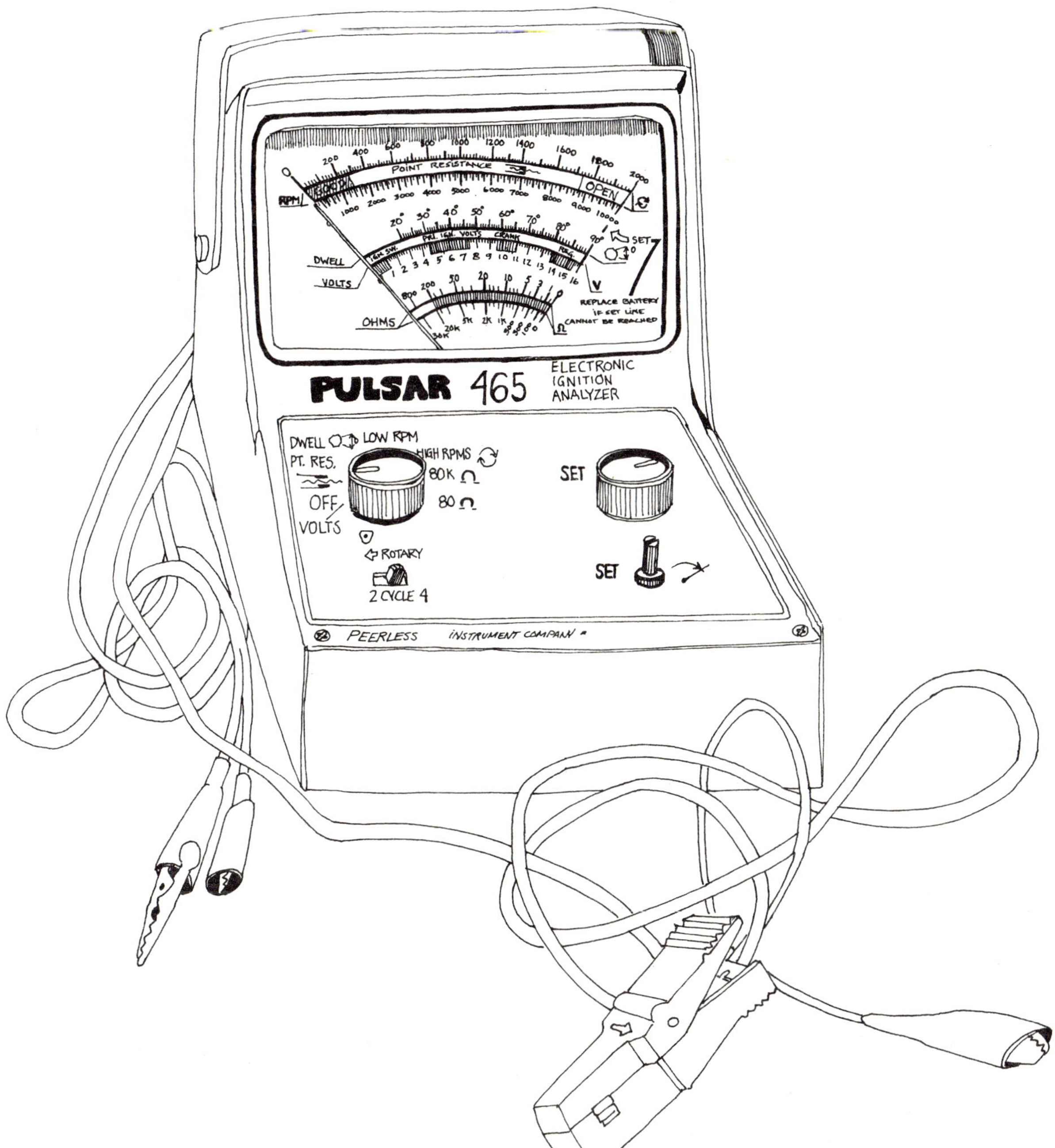

The spark-plug gauge is a collection of calibrated wires with a right-angle bend in the end. Since a .028" wire has a .028" diameter, the gauge is appropriate for measuring the minimum gap in uneven surfaces.

Bonney manufactures a spark-plug gap-setting tool that makes this task go much faster. When I double check with a wire gauge I find that tool quite accurate. It is available through Whitney in Chicago.

This completes the description of the basic tools I have in my tote tray, besides a paper clip, a piece of chalk, a point file, and a few other odds and ends. I will now progress to some additional tools I use.

Other Equipment

Compression Gauge.—Never did I throw my money away faster than when I bought cheap tuneup instruments and in particular compression gauges. Now I use a Peerless screw-in type. It is consistent and easy to use. Since I work alone, it is nice not to have to hold the gauge in the cylinder while somebody turns the ignition switch. You can buy a remote-starter switch, but here again I have had bad luck. Switches made for this purpose have to be heavy duty, but the ones I have used have not lasted long.

Tach-Dwell Meter.—My first tach-dwell meter was purchased from one of those national discount electronics stores. In a short time the meter needed adjustments. I followed the complicated adjustment procedure carefully but it burned up nevertheless. Instruments used in auto mechanics have to be sturdy enough to take some knocking about. It seems to me that some instruments were designed by electrical engineers who had only a schematic-diagram relationship to an automobile. I can recommend Peerless meters.

My favorite test meter is the Peerless Electronic Ignition Analyzer, model 465. It may be used with conventional or electronic ignition; it is a tachometer so it can be used for setting carburetors and for tuning; it is a dwell meter so it measures the setting of the points; as an ohmmeter it can be used to diagnose electronic ignition and electronic fuel-injection, among other things; and as a volt meter it can be used with an ammeter to test the battery, starter motor, generator, alternator, and so on. The ammeter I use is the two-scale induction type meter, also put out by Peerless (model 703). This combination competes favorably with equipment costing ten times as much. Peerless has excellent guarantees on its instruments. The model 465 has two limitations. You have to remember to turn it off when you have finished using it to avoid draining the batteries; and Peerless recommends that the maximum charging rate in a generator system (12) volts) be no greater than 15.4 volts. I recommend it be no greater than 14.8 volts, preferably 14.4 volts.

Timing Light.—Either buy or make a test light and do the timing statically, or get a good strobe timing light. Buy the type that has an induction clamp that clamps right onto the spark-plug wire and gets its power from the car battery. I use a Sears top-of-the-line strobe timing light. They regularly have sales on this item. Neon timing lights are a waste of money because you have to use them in the dark to see anything.

Test Lamp.—This is a simple 12-volt bulb in a fixture that has a probe on one side and a wire and alligator clip for the other lead. It has no batteries. Its purpose is simply to test whether a wire is "hot" or not. It can also be used for timing the distributor.

Continuity Light.—In contrast to a test lamp, this test instrument has batteries and its purpose is to test for breaks in electrical systems. For example, if you wish to know whether a switch is working, attach one lead to the input of the switch, the other to the output, and then operate the switch. If it turns the continuity lamp on and off, the switch works. Testing a circuit that has car battery current running through it will blow the bulb of the continuity tester. Have a few spare bulbs around. Since this lamp has batteries, it must always be tested before you use it. Attach the alligator clip to the probe. If the light comes on, it's alright. If it doesn't, check the batteries and connections. With all battery-operated equipment it is a good idea to remove the batteries if you don't plan to use it for a while.

Alligator Leads.—Two three-foot 16-gauge wires with alligator clips on both ends are useful for making temporary electrical connections. An example would be to "hot wire" the coil from the battery to bypass the ignition switch.

Jumper Cables.—A good set with heavy-gauge copper wire is indispensable. Many people carry them in their car along with a jack and a spare tire.

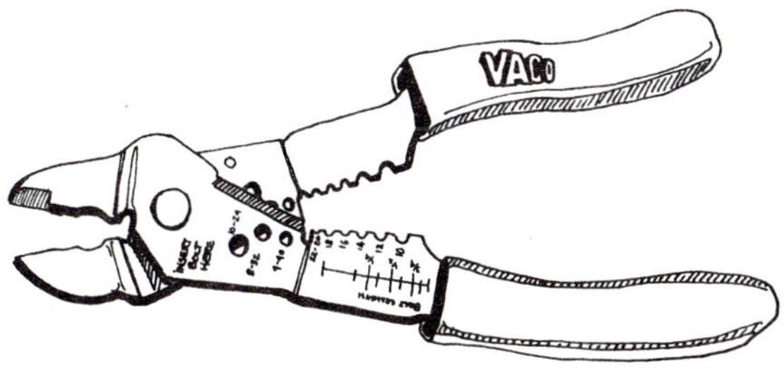

Wire Cutters.—I have both a crimping, stripping, cutting tool and a good pair of diagonal cutters. Soldered connections are still the best, but nowadays everybody seems to use crimp connections. Get a wide assortment of these connectors first. After a while you will find yourself regularly using only four or five types. Have a roll of plastic electrician's tape on hand for wrapping around bare wires.

Oscilloscope.—Although an automotive oscilloscope can be very interesting, I think it is more trouble than it is worth. You can learn to interpret a spark by color and strength as quickly as you can learn to use an oscilloscope. Its other supposed advantages are also questionable, when you consider you are introducing a tool that is very complex, and may need repair and calibration any time. All anomalies that can show up on an oscilloscope because of its own operations would have to be added to the long list of difficulties that are inherently part of solving problems on the automobile. I feel the same about exhaust analyzers. However, perhaps I'm simply not using the right equipment.

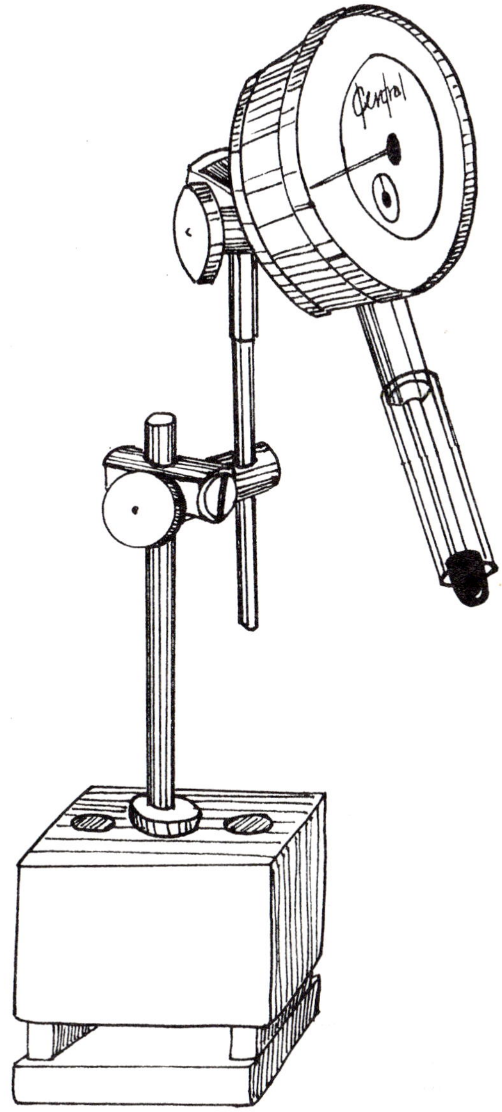

A very valuable measuring tool for engine work and for other problems is the dial indicator. It allows you to measure end play, and "out-of-center" on a shaft, such as the distributor shaft or wheel axle.

A good machinist's steel rule is valuable. In addition to being used for measuring, it may be used to check for warped surfaces by putting it on edge across the object being checked. I have a dial vernier caliper, but I don't use it much. I use Plastigage to measure engine main bearing-journal clearances (see *Engine Overhaul*). I let the machinist do a lot of the measuring for me.

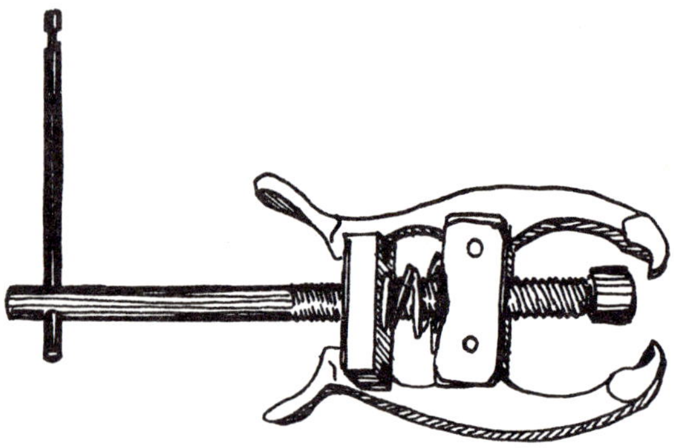

Battery-Clamp Puller.—This is the first puller to purchase. It removes those corroded battery clamps.

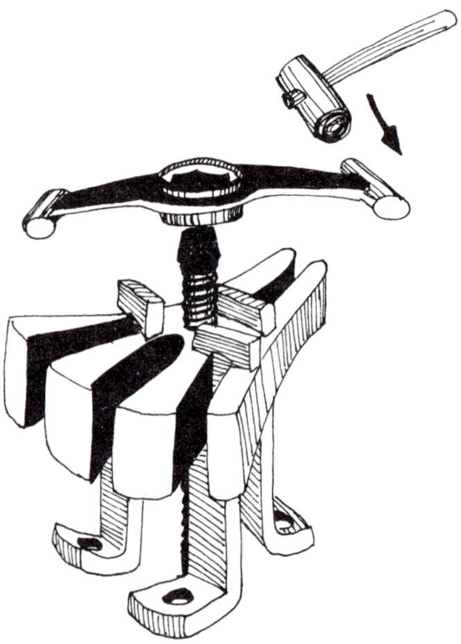

Rear Brake-Drum Puller.—This puller takes off drums that are keyed into a tapered rear axle such as on the Rambler or Volvo. If you are in the business, you will want a universal type that will fit 4- , 5- , or 6-lug drums. Also be sure to get high quality here because at times you have to go at this tool with a sledge hammer. If the parts aren't hardened they will break.

Gear Puller.—Essential in engine work is a puller that can remove parts keyed onto shafts.

Slide-Hammer Axle Puller.—This is for rear axles with studded flanges such as on some General Motors cars. One removes the lug nuts, the wheel, then the drum, which reveals the flange to which the slide hammer attaches. Often a pair of claw hammers prying opposite one another behind the flange will work. If they don't, slide hammers can be rented.

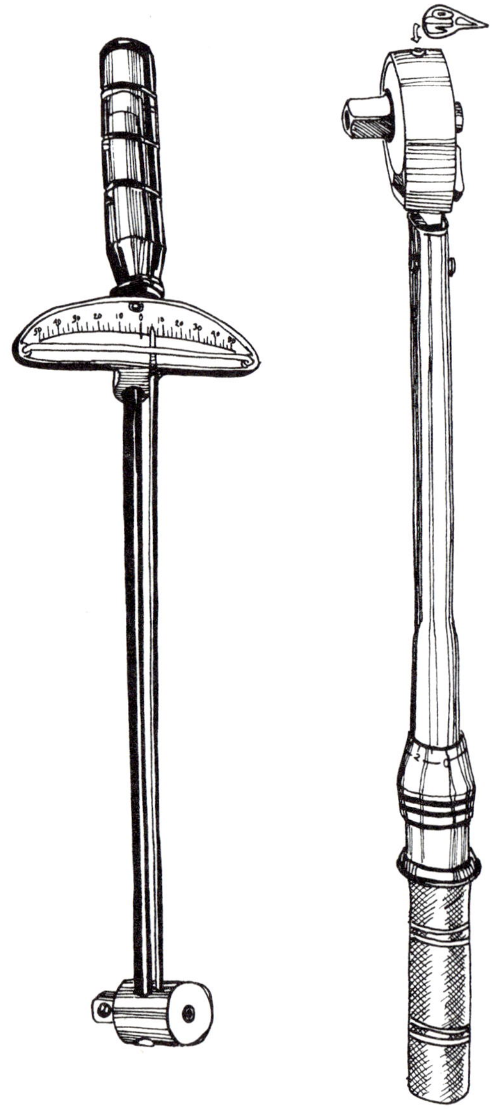

Torque Wrench.—Used for tightening down bolts to a particular tightness (torque), this wrench you need for any major engine work. Ideally you want one that will be operating in the middle 80% of its range for the particular application you have in mind. With cast-iron engine blocks the range will be from 65 lbs. to 105 lbs. torque. With aluminum, the range could be from 30 lbs. to 50 lbs.

There is a beam type that is least expensive and quite accurate. It is a little hard to read when you use it in the higher torques. There is also a spring-loaded type, which one sets to a particular value. When the desired torque is reached the handle clicks. This overcomes the inconvenience of your having to be in position to see the gauge while you tighten the bolts. This type is more expensive, and has the disadvantage of having a spring, which of course will lose its tension somewhat with time. It is best always to leave the setting at the lowest value after use, to keep tension off the spring. There is a third type, a heavy-duty torque wrench with dial, which is even more expensive and about which I know nothing, never having used it. I use a small 3/8"-drive beam torque wrench for torques of 10-50 lbs., and a 1/2"-drive click type for the 40-50 lbs. range. Sears has these on sale frequently.

Knife.—I carry a pocket knife anyway, but if you don't, get one for the tool box. It is good for scraping off gaskets, cutting hoses, removing oxidation from the tip of the rotor, cutting electrical tape, and so on. A $5-to-$10 knife will do; this ain't no bear hunt.

Flashlight.—I use a small plastic one that I can hold in my mouth, after the principle of the miner's hat.

Magnets.—Buy a magnet on a long, flexible handle for picking up dropped tools and parts, and for detecting nuts in long grass. I remember dropping a nut into a Mercedes crankcase while doing a valve job; I was able to fish out the nut through the oil-drum plug hole with a long flexible magnet. I also have a smaller magnet rod for tighter corners, and a long flexible olive picker for nonferrous parts that get dropped.

Flaring Tool.—This tool is for flaring the ends of copper and steel tubing used in gas lines and brake lines. A single flare is all that is necessary for gas lines because the pressure in them ranges from 5 lbs. in conventional setups, to 30 lbs. in fuel-injection systems. Brake lines and connections must withstand thousands of pounds of pressure, and the cost of failure is usually much greater. Therefore, double flaring is required by law in California.

The flaring tools generally available only do single flares, but K-D (K-D Manufacturing Co., Lancaster, Pa.) makes one that does both double and single flares. You probably only need to do flaring work on old cars. Precut flared lengths of brake line are available for contemporary models of American cars.

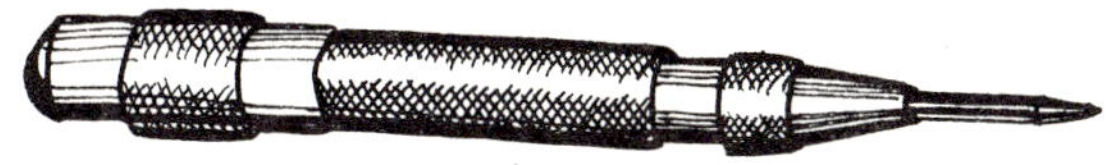

Marking Tool.—This tool is for putting a small depression in metal as a way of marking it. The tool enables one to put marks on rods, pistons, clutch pressure plate, and driveshaft so as to make it possible for you to put them back exactly as they came apart, retaining the factory balance.

Compressed Air.—A compressor is a nice tool to have available. You use it for cleaning parts, especially carburetors and cylinders, when you do a valve job, as well as for filling flat tires and operating air-driven tools. Wear glasses when you clean things, because the air stream will unexpectedly squirt objects into your eyes at high speed. Good portable compressors are available in the second-hand market.

Cold Chisels.—I recommend obtaining a small-size and a medium-size chisel initially and perhaps a diamond-point chisel as well. If you have chisels you should have a grinder to keep them sharp. The "cold" refers to the fact that these chisels are designed to cut or chip cold metal.

Punches.—The prick punch is for marking metal before drilling a hole. Pin punches are for removing pins, naturally enough, but you can probably get away with tapered punches at first, if you are careful not to jam the punch in the hole from which a pin is being driven. Get a big punch for driving out wheel-bearing races from brake drums. Keep the end of the punch ground flat.

Files.—A point file is good for dressing the points in the distributor. I also use a rat-tail file, a small triangular file for dressing up damaged threads on bolts, and a small flat file for taking off rough edges.

320 Emery Paper.—This is the mechanic's sandpaper. It is good for cleaning up bearing journals, lifters, and any other sliding or rolling surface. Removing more metal than 320 emery can remove is best left to the machinist.

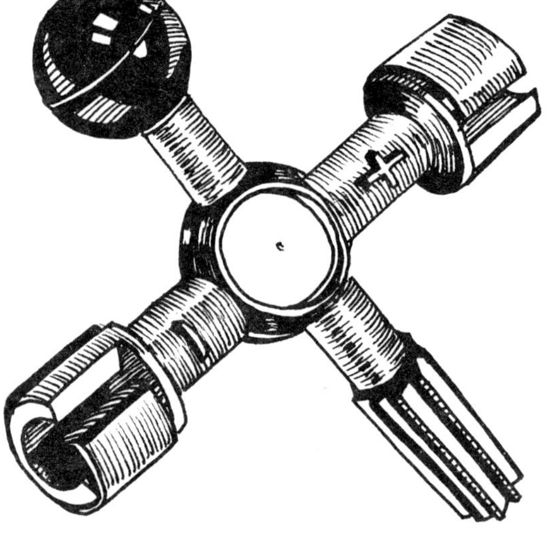

Reamer.—A tapered reamer is useful for enlarging holes a size or two. A special battery post-and-clamp reamer is superior to the wire-brush cleaner. A reamer not only cleans the battery post and clamp but maximizes surface contact between the two.

Bars, Levers, Pipes.—I have a strange collection of these things, picked up over the years, many of them unavailable in stores. I have a 3' x 3/8" pipe for applying pressure on generators or alternators (not too much, please) while tightening the adjustment screw. A 4' x 5/8" pipe to go over my 3/4"-drive breaker bar. I have a rolling wedge bar from Sears that removes seals and goes through screw holes to align two pieces being screwed together. You can also use tapered punches for this. I also have what looks like duck-billed chisels. They have very flattened, tapered, round blades that I use to separate two parts that are stuck together with old gaskets and gasket cement. It is important to do this carefully so that machined mating surfaces are not damaged. I keep old wrist pins and shafts of various kinds to use as drift punches. Even sockets can be used in this way along with a hammer to drive out shafts.

Hammers.—One should avoid using a hammer if possible. A hammer deforms what it strikes. I have four hammers and one sledge hammer. The first is a small brass-headed jeweler's hammer for knocking out small pins and the like. The brass head deforms before steel does, so you do the least damage with this type. Next I have a 12 oz. head ball-peen hammer. I use this with the chisels. The ball-peen part is for taking a curve out of a shaft. This is done by striking the shaft with the ball side around the middle of the shaft on the concave side, contrary to what you might think. I also have a plastic-headed hammer for pounding on things that shouldn't be pounded on. I have to use my 3-lb. hammer on the drum puller and at times I use my 6-lb. sledge hammer for this purpose. Other methods for loosening corroded or rusted parts that should be tried first are penetrating oil applied liberally and let stand for a while, and heat to expand the frozen parts.

Taps and Dies.—These are nice to have. Although the machinist can do the job of rethreading for you, it would usually be inconvenient to have him do it. Of course if you have taps and dies you will also need an electric drill and a full set of twist drills. I find that sizes larger than 1/2" are used most. Therefore, the small set that goes up to 1/2" is a waste of money. Sears often has the large set on sale. Vermont American makes their taps and dies.

Vise.—Purchase as big a vise as you can afford to support. This is a useful tool especially if it has a small anvil section on it. Also it is useful to get soft jaws made of plastic or aluminum that can be attached to the regular jaws of the vise. These may be used when you don't wish to mar the surface of the object you are holding in the vise.

Hacksaw.—It is useful for cutting handbrake cables when they can't be removed any other way, and for numerous other small jobs.

Welding Outfit.—A welding outfit can be useful, even just a jeweler's outfit that will do a small welding job. You have to be quite experienced, however, to do a good job on cast iron or aluminum.

This finishes my general discussion of hand tools used in auto mechanics although it does not exhaust the subject. Very specialized tools will be mentioned in the various repair sections as the need arises. I would like to mention one book on this subject: *Basic Automotive Tools* is an elaboration of this section, giving some basic machine-shop techniques as well as the use of tools. Two catalogs are useful for tool references and, in the first case, tool cost. Sears puts out a *Power and Hand Tool Catalog.* It is a good reference for the cost of a wide range of tools. Also the K-D tool company, which specializes in unique automotive tools, puts out a useful catalog. The essential trick to doing a job can sometimes be learned if one can see the special tool used to do it. Sometimes you can fabricate the tool yourself.

Lifting Tools

Hand tools reflect the refined skills of the auto mechanic, but jacks and hoists reflect the weighty requirements of this trade. A study of the ways to move heavy weights and to keep things from moving when we don't desire it may save damaged tools, crushed fingers, and even lives.

When you are doing just tuneups, carburetor repair, valve jobs, or electrical work there is no need to consider these matters. But as soon as you expand to brake work the car must be lifted. This requires the movement of a thousand or more pounds. Once a weight of this size begins to move, it is hard for a single person to control. Work on as flat a surface as can be found. You can start with only a scissors jack, lifting the car up one end at a time and lowering it back down onto stationary jacks. A scissors jack is weak when it comes to sideways movement—a car can be pushed off a scissors jack from the side quite easily. Before I go under a car I give it a healthy push even if it is on jack stands, as it should be, just to see if it is well balanced.

If a car falls off the jack stands with its wheels off, simply getting it back up again can be embarrassingly difficult. I remember a 1958 limited-edition stainless-steel-top Cad El Dorado whose springs collapsed, which sat on the floor for months before anybody figured out how to move it without damage.

I started in this business in the street with a scissors jack. I'm pleased to say I survived. Now I use a three-ton floor jack, which I purchased from Sears. It is very long, so that it will roll under the longest trunk to reach the differential but, more important, I hardly ever use it at more than half of its lifting capacity. Even so, I still use stationary jack stands if I'm going underneath. Along with the floor jack and a smooth concrete surface goes the creeper, a flat board with offset swivel rollers, which makes life easier. If you are restricted to rough ground, then some large pieces of cardboard placed on the ground make it easier to roll around as well as to find dropped parts.

Another approach used by an older generation of do-it-yourselfers before good hydraulic jacks were generally available, was to build a pit, the proverbial grease pit. I find pits rather cramped and, if not cement lined, subject to rapid deterioration. Besides, pits cannot be moved from place to place; and they collect water.

A limitation of the floor jack is that after lifting the rear end of most American cars by means of the differential, a bumper jack has to be used to lift the body off the springs a few inches in order to get the wheel out of the wheel well. The typical bumper jack is just about the worst jack to use for anything except tire changing, and even for that only if nothing else is available.

Raising a car 20" is usually sufficient to do whatever needs to be done, such as transmission removal. I jack up some small cars just so I don't have to bend over so much. If it takes an extra five minutes to save a backache I'll do it every time. For short jobs on short cars I sometimes use a kneeling stool.

I use a transmission jack, which is nothing more than a heavy-duty scissors jack on wheels. No matter what you use for transmission removal, cursing and bitching are also often necessary to get it back in. The jack helps a lot though.

I also have a small hydraulic jack for numerous little lifting and holding jobs, and a selection of stationary jack stands. Seemingly inconsequential, but useful, is a collection of short boards for wedging, leveling, and raising the height of the jack stands. These just seem to collect over the years.

The next consideration is a hoist for taking engines out. Here is an instance where an old tool might be the best selection. Unless you have electricity available and the money for an electric hoist, a chain fall will do nicely. The come-along is popular these days and it is useful for winching a car out of a ditch. It is even OK for taking an engine out, but because of its ratchet action it drops its load in 1/2" increments. This is quite a shock if you are trying to drop 300 to 500 lbs. gently onto a precise spot.

You have to have a heavy crossbeam, preferably 7½' above the ground, from which to hang the chain fall, and no ordinary garage will handle it without modification. I found some old 4" x 12" beams. One I put under an appropriate roof crossbeam, and jacked it (and the roof) up a few inches. Then I lowered the jack letting the roof pin the crossbeam onto two big vertical supports. I also cut out some panels in the roof and replaced them with corrugated fiberglass panels that gave a diffuse bright white light. Ordinarily the biggest problem in visibility is the contrast between a dark garage and the bright outdoors. Of course the absence of light slows things down. I like to quit when it's dark. This means I work less in the winter, which is appropriate because more energy gets used up just staying warm, and less energy is available.

Another method of hoisting engines out is a hydraulic rig mounted on wheels. This method works all right. Since one can rent all three types of hoists you can try before you buy.

Finally, one needs a couple of four-foot chains with hooks on both ends to act as slings around the engine.

Chemicals

This is my list of folk remedies and super-chemicals that I guarantee to work as advertised.

Solvents.—Gasoline is a marvelous solvent except for one drawback: it is extremely flammable. Especially the fumes. Kerosene is a good grease cutter, but leaves an oily film. Chevron of California sells a solvent called White Pearl that is a compromise between the two. It's a paint thinner. I recommend a solvent you can buy in bulk at the gas station, either kerosene or White Pearl. Gunk concentrate can be added to the kerosene to make a brush-on grease cutter that can be sprayed off with a hose.

For cutting grease and removing brake fluid, trichloroethylene is unbeatable. A Pacific Gas & Electric foreman turned me on to this. I use it to clean up brake parts, and it will save brake-fluid-soaked brake shoes. It is sold commercially and expensively in aerosol cans for brake cleaning. A chemical wholesale house will sell a gallon of the stuff at a quarter to a fifth the cost. It's still expensive, but used carefully it's worth it.

Carburetor cleaner also is much cheaper when purchased by its generic name at the chemical wholesaler. Nothing cuts carburetor varnish like methylene chloride. It can be used over many times. Apply this cleaner with a parts brush being careful about splashing into the eyes. Most cleaning agents, by the by, are rugged on the hands because they remove the oils of the skin. I use a hand cleaner that has lanolin in it, to counteract this.

Lacquer thinner is sometimes the only way to remove old glued-on gaskets. Apply liberally on the old gaskets and let sit for a minute or two, then scrape off. You may have to do it a few times.

Sealers.—Alumaseal is a water-system sealer and a water-pump lubricant that I automatically use after rebuilding an engine. It has very fine particles and is not likely to block smaller radiator passages. There are other similar products.

Siliconsealer was developed by General Electric for the space program. It is sold under other labels, such as Permatex, but they are all the same thing. At present it is cheaper if you buy it in a hardware store as it is just starting to penetrate the automotive market. It can be used instead of a gasket except where gasoline is involved. I use it also to seal leaky windshield gaskets and even to glue wooden partitions into my tool box to organize my tools. When it cures, it remains flexible for years. If you plan to use a large quantity in a short time, it is much cheaper to buy in a large tube in the caulking-compound section of a hardware store. There is also a high-temperature version put out by Permatex.

I use Permatex No. 1 on oil-pan and rocker-arm cover gaskets. One is supposed to glue only the pan or cover side of the gasket, because the block against which it abuts is a machined surface. On oil-pan gaskets where I want to be sure of sealing since there is often no chance to do it over, I sometimes use siliconsealer on both sides of the gasket. There are numerous products for gluing head gaskets. I have used many of them and never had any trouble with any of them. At the moment I use Permatex red because it is so pretty.

Another useful glue is 3M Weatherstrip cement. It is excellent for gluing on weather stripping as well as for gluing together the crank-case halves in a VW. Barge cement, which is used by cobblers, is supposed to be one of the few glues that will withstand the solving action of gasoline. I use fiberglass resin sealers, like those used in body work, for sealing leaky gas tanks.

For anything that seems to come loose all the time or that you want to be absolutely sure doesn't come loose, like connecting rod-cap bolts, main-bearing cap bolts, or flywheel bolts, use Loctite. It is a resin that keeps the threads clogged so the screws won't vibrate loose. There is a double-strength Loctite (red) for freezing bearings into their housings or for motorcycles, which have a bad vibration problem. The Loctite company has a new-formula resin that can be applied after the bolt is in place.

Lubricants.—WD-40 is truly a miracle drug. Often I can solve several problems on an old car by just a liberal dose of WD-40. It can be sprayed on electrical switches and connectors to clean and dehydrate. But it is flammable.

Another problem with it is the method of dispensing. In an aerosol can it works fine, but one would like to dispense with that type of dispenser. Unless one is speaking of something used in very small quantities in a can that would last a year or so, aerosol cans are too expensive in every way to justify their immediate convenience. It's much cheaper to buy WD-40 by the gallon, but the plastic hand sprays you could use deteriorate in a few months with WD-40 in them. Even the hand sprays, which are sold supposedly for this very purpose, are no good in a few months. The alternative is Sure-Shot, a metal sprayer that can be pressurized at your neighborhood gas-station tire pump. Sure-Shot is available from the Milwaukee Sprayer Mfg. Co. (5635 W. Douglas Ave., Milwaukee Wis. 53218).

Anti-Seize compound made by Loctite Company does the opposite of what Loctite does: It lubricates threads where metal galling is a problem. I always use it on spark plugs that are going to be threaded into aluminum heads. Vegas with pop-top aluminum blocks should have Anti-Seize on the head bolts. The additional resistance added by galling between the bolt threads and the aluminum block can easily cause stripped threads in the block or undertorquing of the head bolts. This galling problem can be so bad—especially on some cheap Japanese cars—that bolt heads will strip off while they are being removed.

I use a white assembly grease when I am putting together an engine. It is essential to have the main bearings, rod bearings, and cam bearings lubricated because they will otherwise run dry for the first thirty seconds to a minute the first time the engine is cranked over before the oil pump builds up pressure. My machinist just uses 30-weight engine oil, and this seems to work fine. I like the white grease because it shows the dirt, which I can then avoid. However, while grease or oil will serve as a lubricant when it is clean, it becomes an abrasive when it gets dirty. This is the reason for grease seals around lubricated parts as well as for the oil filter. Of course, the seals also keep the grease from getting into places where grease shouldn't be, like brake shoes and clutches.

Molybdenum disulphide is an additive to a lot of lubricants. It's good stuff. I use a high-temperature disc-brake wheel-bearing grease that has "moly" in it. This grease has the highest drop point of all wheel-bearing greases. The drop point is the temperature at which the grease will liquefy and run out of the bearings onto the shoes. Regular chassis grease will do this at the first sustained breaking.

Disc breaks generate more heat, and therefore require the special grease. To avoid complications I use this wheel-bearing grease for drum-brake systems as well.

I use a dry graphite lubricant called "miracle d.g.f." It's handy where you don't want too much lubrication, as on the transmission input shaft where the clutch disc slides. It can be used effectively on any sliding surfaces. One machinist I know sprays it on heads after doing a valve job to discourage carbon particles from settling on the valve edges.

Finally, it's a good idea to have a box of baking soda around to mix up with water and pour on battery acid spilled or leaked out of the battery.

Small Parts

"Small parts" are items like washers, screws, springs, cotter pins, and the like. If you haven't already started a collection of hardware retrievably arranged in containers such as baby-food jars, then you ought to begin. I recommend that you get a collection of various-size cotter pins, washers, and sheet-metal screws in plastic sectioned boxes at the hardware store. You can get similar collections of external and internal snap rings, and a few universal springs at the car-parts store. Don't get a big selection of springs because these are too specific in their application. After you get these collections you need only replace those of which you use a lot.

A washer allows the free movement of the screw or bolt head as it is tightened, and spreads the area of pressure to increase the strength of the fastening. Obtain a nice selection of these for replacements or additions. Lock washers assist in maintaining tension of the screw to keep it from vibrating loose; they come in several designs.

Ball bearings in inch sizes can be obtained loose from a well-stocked hardware store. Rubber O-rings in standard sizes and standard and metric automotive hardware are obtainable at the car-parts store. When I have to go to the store to buy some particular bolt or nut I usually buy twice or three times as many, so I will be prepared for the next time.

An old box into which one throws odds and ends of bushings, shafts, nuts, and bolts can become a gold mine.

A book that expands on these things is *Mechanical Trades* by Carl A. Nelson. This book is not specifically for the automobile but has good basic information on "V" belts, chain drives, spun gears, couplings, screw threads, mechanical fasteners, packings and seals, bearings and so on.

Books

The book as a secondary tool is so important that I was able to teach myself the repair trade virtually without any apprenticeship to a mechanic. Alternating between a book and the tools I was able in three years to reach an hourly pay level above that which a shop mechanic makes. Along the way I acquired quite a library. No individual book covers the subject comprehensively. The aim of this section is to indicate some books that are useful, and for what.

A Manual of Step by Step Procedures, published by John Muir Publications, is a fairly comprehensive book if you want to limit yourself to VW repair, and don't care much for theory. The VW is a good car for a first crack at engine rebuilding because one can get away with fairly simple tools and the book provides clear step-by-step instructions. However, I would ignore Muir's bootstrap method of measuring crankshaft end play, and buy a dial indicator. Muir has a complete description of the vagaries encountered when valves are set. His is one of the few books that has the smell of grease. Most of the rest are as dry as sticks, nearly unreadable. The book's major drawbacks are poor indexing (an aliment of most automotive books), a shortage of pictures, and no electrical-system schematics. Also, for reasons unfathomable to me, the author considers carburetor repair more difficult than engine overhauling, and so refers you to an "expert." The VW carburetor is about the simplest around.

Chilton's Auto Repair Manual and *Motor's Auto Repair Manual* are the two standard bibles of auto information for the general repairman of American cars. They are similar in that they are published annually and summarize the information from shop manuals for the preceding decade. Like the telephone book they have a new edition each year. For any given car the information is very bare bones indeed. A lot is assumed of the reader. The books are good sources for tune-up data, and provide a quick review of the steps necessary for any normal auto-part removal and replacement (R and R). As other sources, these manuals have no electrical-wiring diagrams. Maybe they think mechanics cannot read them. Actually, hardly any description conveys as much information in such a small space as a wiring diagram. European car manuals usually have wiring diagrams.

Motor's manual is usually sold through the mail to gas stations, whereas *Chilton's* is sold in parts stores. I have both books in ten-year intervals from 1935. Together they offer a more complete picture of the subject because they abstract the shop manuals in a slightly different way. The older copies are available in second-hand book stores and are not particularly cheap. The older series of *Chilton's* on foreign cars was and is terrible. There is now, however, a two-volume *Chilton's* in large format, on foreign cars, that is much improved, although a new edition is needed. A very fine series of workshop manuals is available on nearly all foreign cars by *Autopress*. I would get one of these for any major overhaul.

The book that you are reading plus an *Autopress* manual would be sufficient for major engine work on a foreign car, and this book plus a *Motor's* or *Chilton's* would be sufficient for more straightforward American-car work. At least they would be sufficient as far as books go.

The publisher Motor in New York puts out *Auto Engines and Electrical Systems*. For me it is a bible for basic engine theory. It's like having a first-class machinist on the bookshelf. The book is a very detailed examination of what goes wrong with engines, how to recognize it, and how to fix it. Motor also publishes a *Truck Repair Manual*.

Petersen puts out a series of books called *Basic Automotive Tools, Basic Carburetion and Fuel Systems*, and others. I recommend the first as a supplement to my book, and the rest if you are interested in high performance, or hotrod modifications.

There are many books in the following two categories: textbooks, which are somewhat distant from the dirt and grease of actual shop practice, and do-it-yourself books, which treat difficult subjects too simplistically. One other book I want to mention, even though it is out of print, is Colin Campbell, *The Sports Car*. It's a little dated, but a fine, readable book nevertheless. I should say in closing that every book I have picked up on the subject has taught me something. Even the car owner's handbook has a few pieces of information in it not available elsewhere.

Odds and Ends

Rags.—Lint-free cotton rags are the best. Baby diapers are extremely absorbent although not completely lint free. Artificial fibers are worthless except for wiping grease off your screwdriver. Dirty rags should be kept in a closed metal container.

First Aid.—I will not discuss the first aid necessary if a car drops on you, but rather the everyday problems faced by the mechanic, which may be reduced with care but not completely eliminated. If a carpenter's problem is wood slivers, a mechanic's problem is metal slivers. Therefore a good pair of tweezers are handy to have nearby. Also an eye cup to clean brake fluid or other caustic agents from one's eye, and a mirror to find that speck of dirt. Cuts and abrasions are nicely healed by the petroleum products that get on one's hands. The raw petroleum that seeped from the ground before the days of the automobile was sold for many medicinal purposes. The *Merck Index* lists petroleum as an antiseptic. In other words, if you skin your knuckle while you work, just get a little grease on it and keep going.

Fire Extinguisher.—Have one or two around that are designed for putting out electrical or petroleum fires.

Locks and Alarms.—I have had my tools stolen twice. I now have a fenced-in work area, and locked boxes with alarms on them, which sound if the locks are broken. To the common thief, tools are as desirable as televisions. Therefore, I'm sorry to say, a little foresighted paranoia is in order. I've had strange people in Cadillacs ask me what I was in the market for—they would get it. I advise against the purchase of hot merchandise because it helps to stimulate a business that *most* of the time is working against your best interest. In fact, hot merchandise has a way of disappearing again.

Now it's time to pick up the tools, the books, and the rags, and move into the shop.

Chapter 3:

Tune-up

The purest experiment in treatment may still be conscientious. . . . Science is properly more scrupulous than dogma. Dogma gives a charter to mistake, but the very breath of science is contest with mistake and must keep the conscience alive.

—George Eliot (1819-1880)

When I get a "new" used car I do a complete by-the-book tune-up and replace all typically replaced parts such as plugs, points, condensor, rotor, and cap. I adjust valves, check compression, set the carburetor, and rebuild it if necessary. I do this to set a base line for performance. I want to know what the car can do. After that I ignore the routine tune-up procedure, but I listen to and feel the car and do what needs to be done when it needs it. On one occasion I may only adjust the valves, on another only set the points. This pragmatic approach is probably not suited for the ordinary non-mechanic, but it is obvious that the regular tune-up routine suggested by the manufacturer is only a rough approximation designed to catch parts just before they wear out. Interestingly, I find musicians, as a group, the most sensitive to their cars being in or out of tune, perhaps because they can listen.

For the average driver the best measure of over-all performance is gas mileage. If records are kept consistently, they will tell you whether something is wrong, although they won't tell you what is wrong. A tune-up is usually the remedy.

The tune-up is the equivalent of the doctor's general checkup. It gets you into the engine and even solves some of its minor ailments. These minor ailments are ususally discovered in the conventional ignition system rather than the carburetion system. The tune-up as described here is primarily a maintenance procedure, which will reveal what other work may have to be done on the engine. With a new customer I generally take a short case history on the car. I can often recognize a problem better if I have the car's age and an idea of its deficiencies; never make a judgment about a car's state of tuning when the engine is cold.

Spark Plugs

The first step in a tune-up is the removal of the spark plugs. Since most cars now use carbon-core plug wires to cut down on radio-frequency noise produced by the ignition system, it is essential to remove the wires from the plugs carefully. The cores are vulnerable to internal breaks, so pull by the boot when you remove the wires from the plug. The boot should be a snug fit. Place the wires in such a way that by their position, length, and shape they will remind you on which plugs they go.

As the plugs are removed with the special spark-plug socket, line them up in the order you are removing them so you can tell which plug came from which cylinder. Some plugs are very difficult to get out. The anterior plugs of the older Porsche cars are the worst. In this case you will have to work blind. Take out one of the easier ones first, to get an idea of the angle of the plug in the engine. Try different combinations of ratchet, short extension, and universal joint with the socket. Expect to do some strange contortions to enable you to get your hand around the carburetor and in the right position to apply leverage to the ratchet handle. On the 1700 and 1800 VWs, note the angle at which the plug enters the head when you remove the plug. You will have to work under the exhaust manifold on many V-8s, which, if the car has just been driven in, will be as hot as a pistol.

Examine each plug, comparing it with a new one. I generally find that I can reuse a car's spark plugs about half the time. Make sure the center electrode is square-tipped or very nearly square-tipped and that the side electrode is not burned. A light coating of deposits can be scraped off carefully with a pocket knife, and the plug regapped with a Bonney gapping tool. I have checked the Bonney tool with a wire gauge and find it quite acceptable.

You might keep around a few new or reasonably good used plugs of the types you see a lot, for emergencies, such as when you accidentally break a plug upon removal. It's a shame to have to go to the parts store for one plug. Parts-store trips can consume a lot of your time.

If the plugs show a very light tan or gray, the combustion situation is healthy. If a plug is wet, smell it. If it smells like gas, it isn't firing either because of very low compression, a bad wire, or a bad plug. If all plugs are wet, there is probably no spark from the coil, or perhaps the carburetor choke is stuck. If there are black flaky deposits on the plugs, there may be carburetor problems. Heavy deposits may suggest the need for a hotter plug, particularly if the car is only used for around-town trips. I had a real-estate-agent customer who did much low-speed driving. The plugs on her car always had heavy deposits, but the engine was otherwise in excellent shape. A hotter running plug will tend to burn off the deposits. Each plug type comes in a range of different heats. A hotter plug would be used for around-town driving whereas a cooler plug would be better for freeway and desert travel. The absence of deposits on the plug does not eliminate the possibility of bad oil-scraper rings. I remember an old Chevy Six burning a quart with a tank of gas that nevertheless had a clean-looking combustion chamber and plugs.

In multiple-carburetor setups the plug color is often useful to reveal which carb is set too rich or too lean. If the rear-carb choke is stuck open, for example, then the plug color might range from light tan to black as one moves from number-one plug to number-four plug.

Fresh oil on the plug could be due to a broken ring, which might have been caused by the timing having been set too advanced. Associated with this would be a severe pinging sound, caused by the pistons being banged around in the cylinders by an early explosion. This sound is most pronounced at low engine speeds under a load (up a hill for example). In situations where cars are run very hard and hot, as in racing, you might see melted electrodes and other strange plug signs, but these are rarely encountered in the average car.

The color of the inside of the tailpipe will be similar to the color of the plugs. A look at the color is a simple quick check to get an idea of the nature of the burn in the cylinders. Black tailpipes indicate unburned hydrocarbons. The situation is similar to that in a fireplace. Fast hot-burning pitchy wood, like pine, leaves the back of the fireplace black. A slow-burning hardwood, like oak, leaves the back of the fireplace gray. The Champion spark-plug people make an informative color chart of plug conditions, which your parts-store might get for you. I prefer to use original-equipment spark plugs. If you don't specify, the parts man will probably give you Champion plugs.

Compression Test

I always take a compression test at this point. If one gets 50 lbs. or less compression in any cylinder, there is no point in continuing the tune-up. I remember two Fords I had at the same time, both of which had a similar symptom, a rough idle. The first car had excellent compression, and the problem turned out to be a dirty carburetor. The second had 30 lbs. in one cylinder, and only a valve job cured the problem. If you continue the tune-up on a bad engine, you waste your time and the customer's money. You are likely to give back a car running worse than when it came in. I charge for checking the compression and let the customer decide whether he wants to embark on major engine work.

I screw in my Peerless compression gauge into the spark plug hole. Then I turn the engine over eight times for each cylinder with the ignition key. Two things will affect the absolute reading on the gauge. If the starter motor is not turning the engine over at a normal rate, the readings will be lower; and if you don't take all plugs out, the reading will first be lower because the engine will be working against the compression in the other cylinders. Again, it is the relative reading that is crucial, therefore take readings for each cylinder in the same way.

A low compression can be due to bad rings, a blown head gasket, or bad valves. The rings can be temporarily sealed if you squirt some engine oil into the cylinder and take the compression once more. If the compression goes up, then the rings are probably bad. If there is no change, then valves or head gasket are to blame. Two adjacent cylinders with a low compression suggest a blown head gasket. Water in a cylinder also indicates a blown head gasket (see *Valve*).

The books say that it is a good sign if the cylinders are all within 10 lbs. of one another. When the difference, even if only in one cylinder, is greater than 50 lbs., it's going to be an expensive job. On a VW, a compression reading of 100 lbs. or less in any cylinder means trouble soon. And VWs tend to suck a valve or blow a piston before they show deterioration signs obvious to the average driver.

It is the balance in an engine that ensures long life and good performance. It is the vibration created by imbalance that kills machinery. I had an old flat-head Ford that had only 70 lbs. in the cylinders, but 70 lbs. in *all* cylinders. In five years of constant use the only engine work I did on it was to replace a blown head gasket. Also there was a 356c Porsche I maintained for a few years that had 70 lbs. in all cylinders. It burned a little oil, but it could take any stock 1600 VW with ease. In fact Porsche doesn't even list a "normal" compression for its cars.

There comes a point where the over-all compression is too low. This may be indicated by an unacceptable level of oil consumption, but also by a hard-starting problem, with the engine spinning but not catching. The heat that causes the explosion in the cylinder comes from three sources: the spark plug, the residual heat in the cylinder walls, piston, valves, and so on, and the heat created when gas is rapidly compressed. Since the cylinder is cool on a cold start, the heat of compression is a crucial part of the heat necessary for the first explosion. A car with a low overall compression might very well run acceptably when it is warm, but not be able to start when cold. In a diesel engine with at least a 15-to-1 compression ratio the heat of compression contributes about 1000°C to the total heat. This extra heat allows the diesel engine to burn the less combustible diesel fuel.

On the other side of the coin one can get ignition without a spark-plug firing, if the engine is running too hot. Residual glowing carbon particles can ignite the mixture. Dieseling, or running-on after the ignition key is turned off, occurs because the heat of compression causes ignition without the aid of spark plugs. The ignition occurs too early, tending to cause the engine to turn backward. This problem can be difficult to solve sometimes. Even car manufacturers have found the problem so intractable that they have invented devices that go on the carburetor, which shut off the idle jet when the ignition is turned off.

If one records a couple of zero-compression readings on what otherwise seems to be a fairly healthy engine, it's time to check the compression gauge. Probably the tire valve in the end of the tool is malfunctioning. It's a good idea to have a few spares around, as well as a few replacement O-rings. Another problem I've had with my compression gauge is that occasionally a piece of dirt will get stuck in the tire valve. If this happens, the pointer on the gauge will be falling instead of remaining steady after the engine is cranked. Usually a little WD-40 spraying in and around the valve, as well as exhausting the gauge through the end valve rather than the side valve, cures the problem. Also, if one hears a hissing noise when the engine is cranked, probably the screw-in part isn't seated properly. Unscrew the gauge, clean and oil the threads, and start again.

I would like to say that if you get a high, even compression your engine is hunky dory. Unfortunately a compression check doesn't tell you anything about the bottom end of the engine, the rods, crank, or cam (see *Engine Overhaul*). It can even give you high readings when in fact your oil-scraper rings are nearly shot. Therefore only negative results are really informative.

Spark-Plug Wires

Leave the spark plugs out until you finish with the static timing, which comes later. This allows you to turn the engine over more easily. Now you move backward through the ignition system along the wires to the distributor cap.

If the wires are oil covered, old, and the boots cracked, consider replacing the wires. They don't cost much. They have to carry 10,000 volts, a voltage that can find many other ways to go to ground rather than to jump the spark-plug gap. This is a high voltage, but the amperage is low and therefore not directly dangerous. (The only direct injury one can get from a car's electrical system is a burn from crossing the large battery cables, especially if one is wearing a ring.)

Be sure all high-tension spark-plug cables are firmly attached to the plugs and, at the other end, firmly seated down into the distributor-cap receptacles. If they are not, the spark plug will jump across the space created by the loose connection with a loss of voltage potential across the spark-plug gap. Also, the wire and the receptacle will become badly burned. Inspect the high-tension wire from the center of the coil to the center of the distributor cap with the same possibilities in mind. On a Lucas system, the high-tension cable that connects the coil with the distributor screws into the coil.

If you have an ohmmeter, you can test these high-tension cables. Wire-core cables should measure no resistance. Most cars these days have resistance wiring, which cuts down on the radio-frequency noise produced by the ignition system. It cuts down on performance as well. The resistance readings on this type of electronic suppression wiring should not exceed 20,000 ohms (20K ohms) by very much. If it does, replace it.

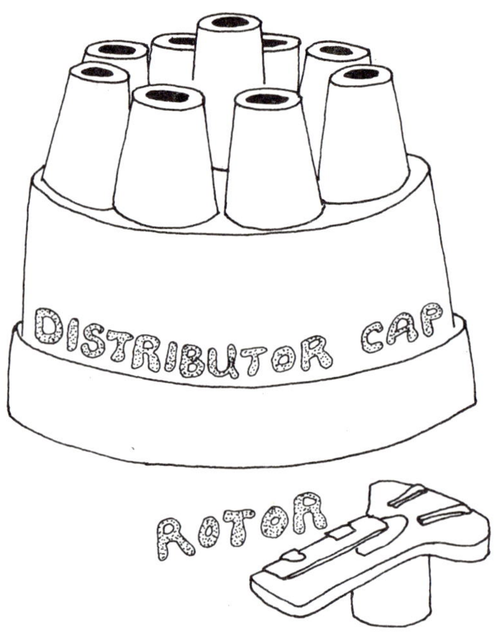

Distributor Cap and Rotor

The distributor cap and rotor distribute the high voltage discharged from the coil to the various spark plugs in the appropriate firing order. The main potential problem with the rotor and cap is a buildup of an oxide on the copper terminals inside the cap and on the tip of the rotor. This oxide is nonconductive, so it should be removed. Take a knife and scrape off the deposits down to, but not including, the metal. You don't want to enlarge the gap between the rotor tip and the cap terminals. If the terminals are too eroded or gouged, replace the cap. Also look for black, spidery lines indicating a spark jumping around inside the cap. The car will miss and backfire if this is happening. Look for corrosion in the wire receptacles on the top of the cap. This can be cleaned out with steel wool.

When you replace the cap or rotor, take the old rotor with you to the parts store. Compare it with the new one brought out by the parts man. The parts business is so crazy that parts comparison is not just a good idea but essential. You won't insult a parts man by doing this. It's not uncommon to find that the car on which you are working has parts that differ from those it is supposed to have according to the book. If the rotor is correct then the cap will be OK.

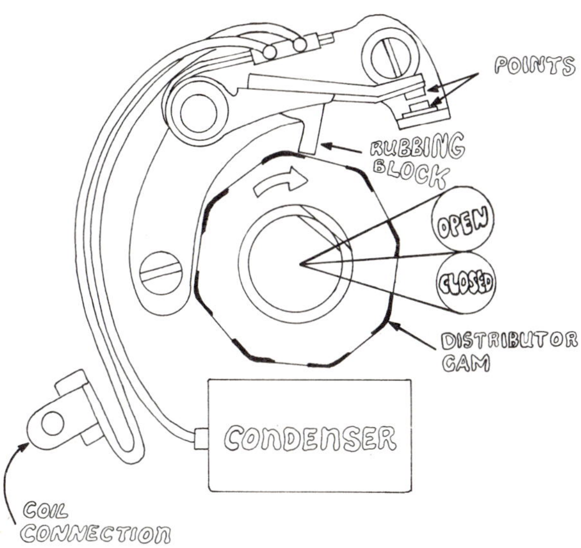

Distributor

Sometimes I remove the distributor, leaving the cap and wires in place, but I don't recommend it until one is thoroughly familiar with basic timing and how not to lose it (see *Engine Overhaul*). This is safe to do with Bosch distributors because they only go back one way.

Look at the breaker plate (the plate on which the points are mounted), and look also at the condensor. If you find copper dust, there is probably a defective cap and rotor, and perhaps a bent distributor shaft. A blackened circular area around the points indicates arcing. Replace points and condensor. Gently separate the points and observe whether they are badly burned. If they are, replace them with the condenser. If you intend to use a feeler gauge and set the points by gapping, replace the points unless they look new. A little tit on one of the used points will be enough to prevent an accurate setting.

Observe whether the rubbing block appears damaged, whether there are loose or burned wires, and whether the contacts are parallel and opposing one another. If any of these is out of order replace the points. Actually, one can realign points, but only by bending the stationary member of the pair.

If you use a dwell meter to set the points, then slight pitting is acceptable as long as the dwell reading is correct. I actually set the points by eye (about the thickness of a matchbook cover), then measure with the dwell meter later when the engine is started for the final stages of the tune-up. Resetting is often necessary, but by now I find my eye to be as accurate at estimating dwell as a feeler gauge. Setting by dwell—which is the number of degrees of rotation during which the points are closed for each cylinder—is more accurate than a feeler gauge, because it is an electrical measurement of an electrical function. Setting points with a feeler gauge is only a way of approximating this setting.

Condenser

Before we discuss the installation and setting of the points we should consider the condenser. A bad condenser is rarely encountered but when it happens it might be difficult to diagnose. With a major overhaul I do it. On a Fiat 128 I do it because the condensers are situated right above the exhaust manifold, and sometimes I do it out of pure superstition.

A bad condenser means no spark, and therefore the car will not start. Before you suspect the condenser, make sure that the coil is getting current, that the points aren't grounded, and that all wires connecting points, condenser, and coil are alright. Then check the condenser by replacing it. Any condenser you have around will do for a check.

One of my more difficult problems has been loose condensers, not bad condensers. The screw that holds the condenser also provides the necessary ground between it and the distributor. A loose condenser can mimic a gas-delivery, rotor, or distributor-cap problem.

Before replacing the points and condenser I spray the distributor with WD-40, then wipe off the excess. Then I put a small dab of distributor-cam grease on the distributor cam. This is a special, very high-temperature grease. Then I put a little engine oil on the wick in the center of the distributor shaft if it has one. I rotate the breaker plate. I check if it moves a short distance and springs back. If there is a vacuum advance, I rotate the breaker plate, place a finger over the vacuum input connection, disconnected from the carburetor, and check if it holds a vacuum. If the breaker plate simply springs back to its normal position, the diaphragm is perforated and a new vacuum unit is needed.

Another approach to checking the action of the advance mechanisms is to observe the movement of the timing marks with a strobe light at a later stage in the tune-up.

We are at the half-way point in the tune-up. This is when I go to the parts store to get the new plugs, cap, rotor, points, condenser, wires, and air cleaner, hoping that I have not overlooked anything. Take with you the used parts and know the year, make, and model of the car. If you go to a cut-rate store, be on the ball. They may sell you anything (more in *Doing Business*).

Install the condenser tightly, and the points so that they are firm but movable. They have to be adjusted. When you set the points by eye or with a feeler gauge, the rubbing block has to be on one of the noses of the distributor-shaft cam. This means the crankshaft has to be turned so as to turn the distributor shaft. There are several ways to do this, each of which is appropriate for some cars and not for others.

One way is to hot-wire the starter solenoid. First find the starter solenoid. It has the hot (ungrounded) battery cable running to it. Then find the small-gauge wire that goes to this solenoid switch. This is the ignition-switch wire that closes the solenoid and operates the starter motor. If the car has electronic ignition or electronic fuel injection, disconnect all wires coming to this small terminal. Take an alligator-clip lead and attach one end to the hot cable from the battery. Momentarily touch the other end to the post where the mentioned wire is attached. This will jog the engine a bit. Keep doing this until the rubbing block of the points is sitting on a high spot (cam nose). On cars with Lucas electrical equipment the starter solenoid has a mechanical push button which allows you to do the same thing more easily. Ford systems, which have a separate solenoid, require you to hot-wire the S-terminal of the solenoid. In passing, if you wonder whether you actually have a hot wire in your hand, touch it to some unpainted metal. If it sparks, it's hot.

Another method of turning the distributor shaft is to turn the engine over using the fan belt. Also, if the car has a standard transmission, then putting it into top gear and rolling it will rotate the distributor shaft. This is the only thing to do to a Sprite or an MGA.

Setting the point gap is the first step in setting the timing. Timing refers to the moment when, in the cycle for each cylinder, the spark occurs. A timing of top dead center (TDC) means the spark is set to occur just as the piston reaches the top of the compression stroke.

Changing the point gap changes the timing. The wider the gap the earlier in the cycle the points open. Therefore, distributor timing has to be reset if the point setting is changed. The points are an electrical switch, different from those with which we are more familiar, in that they cause an electrical discharge of the coil upon *opening*. The period during which they are closed (the dwell) is the period when the coil is charging.

To increase the period during which the points are closed (dwell), one decreases the size of the gap.

We want the points to stay closed for a long time to allow the coil to charge as much as possible; at higher engine speeds this time gets shorter. But if we increase the dwell (decrease the gap) too much, then the rubbing block doesn't move smoothly up the ramp of the distributor-shaft cam. Rather it gets blipped by the nose of the cam, causing point bounce that also limits high-speed performance. The dual point, dual-coil system, used in a Maserati, is an attempt to cope with this. The points and coils alternate with one another allowing twice as much time for each coil to recharge.

Cars like the Datsun and Fiat use dual points, but these are for emission control. The primary points are set normally. If you don't know which are the primary points, disconnect one set and try to start the car. If it doesn't start, you've found the primary points. To set the secondary points, either do it with a feeler gauge or by temporarily connecting them up with the primary lead. Set the dwell, then reconnect in the proper fashion.

Cars using Bosch electronic fuel-injection systems have a terminal on the distributor to measure engine speed and to trigger injections. Leave this alone, unless you replace it as a unit (see Fuel Delivery III).

Coil

Many Delco systems on V-8s are adjusted with dwell meter and an Allen wrench from the outside of the distributor while the engine is running. A nice way to do it.

The coil works in concert with the points and condenser. To check the operation of that trio and the wires that bind them, pull the fat wire out of the center of the distributor cap. If the car has an older Lucas system, the distributor cap will have to be taken off, then a small screw removed from near the center, inside the cap, before the wire can be removed. With the distributor cap removed to expose the points, and with the ignition on, gently open and close the points with a pencil or a small stick. Hold the fat high-tension wire with a clothes pin ¼" away from a clean part of the engine and observe the spark. A weak coil will give a weak, narrow, reddish spark or no spark, and a strong coil will give a broader, whitish spark, which can jump a greater gap. If you had an oscilloscope you could measure the peak voltage of the coil discharge. This would also tell you the health of the coil, but by the time you learned how to use and interpret an oscilloscope you could have learned to interpret a spark by eye. To teach your eye the variations in spark, experiment with coils of known quality. And look at the spark produced by a high-energy coil like the Bosch Blue Coil. A weak coil will run hot, and missing will occur at high speeds.

A coil can be wired backward and still work, but with 15% less output, so check that. The small coil-to-distributor wire will either be marked as distributor, or on a Bosch coil, with the number one. If the terminals are marked plus and minus, then the coil-to-distributor should go to the sign (+ or –), which is the same as the ground sign of the battery: negative on most cars, positive on old Fords and most English cars. The other small wire goes to the ignition. If you wish to bypass the ignition switch to be sure you are getting juice to the coil, put a clip lead on the hot battery terminal and the other lead on the ignition terminal of the coil. For more on coil trouble shooting, including the condition of no spark, see *Electrical Systems II.* Do not hot-wire the coil on cars with electronic ignition.

Electronic Ignition

Electronic ignition is an improvement over the traditional system of breaker points, condenser, and coil. It provides higher voltage at the spark plug for quicker starting and better performance at high r.p.m. It cuts down on plug fouling, increases plug life (to 50,000 miles or more) and point life if these are used.

Its weakness is that short circuits or voltage surges caused by conventional under-the-hood short cuts such as hot-wiring, or careless use of jumper cables, will destroy the electric components in the electronic ignition. Don't jump start or run the car with the battery disconnected. Don't remove electrical components for replacement unless the ground strap is disconnected from the battery first.

Normal tune-up maintenance consists of simply seeing if all connections are clean and tight, and checking the points on those systems which use points, such as the first Ford systems.

Set the points at .020" on the early Ford electronic ignition unless otherwise indicated. The dwell can be measured by first connecting the black test lead to the large black terminal of the tach block, and connecting the red test lead to the small red terminal of the tach block. The tach block is mounted between a cold-start relay and a resistor block. Pull the high-tension wire from the center of the distributor cap and ground it with an alligator-clip lead. With the ignition key on, hot-wire the S-terminal of the starter solenoid to crank the engine. The dwell should be 24 degrees from the V-8 engine. The dwell is predetermined by the control box in most electronic ignitions and is not adjustable.

Checking for a spark in an electronic ignition may be done as follows. Remove a spark-plug wire from one of the spark plugs rather than removing the coil high-tension wire from the distributor as was done with the conventional ignition system. Take any good spark plug, stick it into the connector of the plug wire, and then ground the threaded base of the plug to the engine. Now crank the engine, using the ignition key, and observe whether a spark jumps the gap of the spark plug. If it does, then the electronic parts of the ignition are OK. If there is a problem with the engine surging, it may be due to a lean carburetor or an intake-manifold leak, or to faulty grounding between the electronics box and the body. Further trouble-shooting of this type of ignition, especially the no-spark condition, will be covered in *Electrical Systems III.*

Now back to starting the car. If it starts but runs rough, the plug wires may be confused. On a V-8 you can reverse two wires and the car will still run moderately well. On a four-cylinder car the problem is more obvious. Chilton's has a diagram showing the correct wiring pattern.

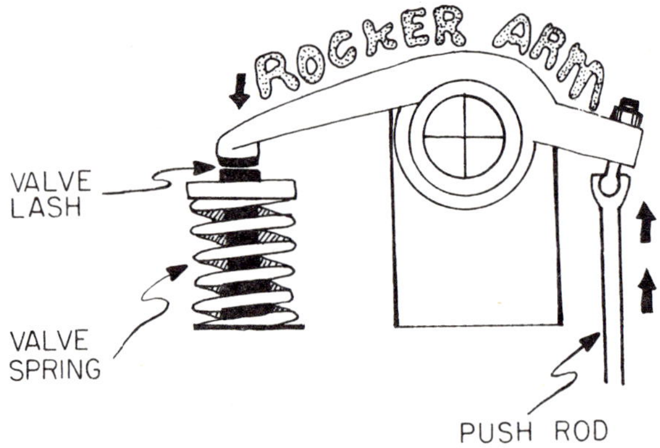

Valve Adjustment

Since we are at the point where we would like to start the car, we have to consider the valve adjustment. On a VW I would adjust the valves with every tune-up. On other cars with solid lifters or overhead cams, perhaps every other time, unless the car sounds like a sewing machine or unless a dealer just tuned it. The sewing-machine noise is due to excessive valve clearance. If an engine is supposed to have a valve clearance of .015" then crankcase expansion may remove as much as .010" of the clearance with the engine running at high r.p.m. under load, or during a hill-climbing exercise. Some manufacturers recommend adding .002" to the standard clearance, if high-speed motoring is anticipated. Hydraulic lifter valves generally don't need adjustment.

Every manufacturer has its own special formula for valve adjustment. Look in Chilton's tune-up section. Some cars have their valves adjusted hot, and some cold.

Most valves on push-rod or overhead-cam engines with solid lifters are adjusted by loosening a lock nut and turning an adjustment screw. Engines with hydraulic lifters are often adjusted by simply turning a self-locking nut. With the engine idling, back off the adjustment nut until the valve begins to clack. Then slowly turn the nut back down until the clack stops;

then make an additional 1/2 turn or whatever is recommended. You may have to make a dam out of tinfoil to keep oil from spilling all over the place. Some overhead-cam engines are adjusted if the shims on the cam are changed. Adjustment on this type is more time consuming, but generally not necessary until the head is removed for an overhaul. For some cars a special tool is used to depress the valve so the shim can be removed and placed by one of more appropriate thickness or achieve the desired valve lash. (For more information on these types see *Valve Job* .)

If you have just overhauled an engine and wish to set the valves before starting the car, and they give you only the hot setting, just add .004" and you'll probably be close. The exhaust valves always require the larger clearance of the two values given, as they get hotter.

To discover which valve is the intake and which is the exhaust, follow on a straight line across to the place where the manifolds bolt to the head. An intake tube will be opposite an intake valve and an exhaust tube will be opposite an exhaust valve. Do not set the valves when they are open, but rather when the cam is on the round side for each cylinder and pair of valves. In an overhead cam you can see this directly. In other engine types, you rotate the crank so the rotor points to a plug wire. This positions the cam correctly for setting the valves of that particular cylinder; then you rotate the crank into position for the next pair of valves, and so on.

Half the time, tightening the lock nut on the adjustment screw changes the adjustment because varnish has made the threads sticky. This may require some finagling, such as setting the adjustment loose in anticipation of this overtightening tendency. In bad cases the adjustment screw and lock nut should be replaced. I always use a go-no-go feeler gauge and recheck the setting after the lock nut is tight.

The rest of the clearance, besides that introduced for metal expansion, introduces a performance factor. The looser the valves are set, the shorter will be the time they are open and the less will be the distance they open to the point of not opening at all. In high-performance engines the design of the cam is such that the valves are made to open farther and for a longer time. This increases the volumetric efficiency of the piston cylinder, which simply means that more fuel is packed into the cylinder on the intake stroke of the piston.

In valve timing it is necessary that the intake valve be open at least at top dead center just as the piston begins its downward intake stroke. Usually opening is planned to occur 5 or 15 degrees before T.D.C. The intake valve remains open for approximately 60 degrees after the piston reaches the bottom of the intake stroke. The reason for this is to take full advantage of the vacuum-created fuel inrush before the compression upstroke begins to take its full effect.

The exhaust valve opens 45 degrees before the bottom of the power stroke, but after most of the effect of the explosion has made its impression on the piston. The valve should be open as long as possible without lessening the efficiency of the power stroke, so as to maximize the removal of exhaust gases which include carbon dioxide. Carbon dioxide retards burning. The valve remains open for 5 to 10 degrees after reaching the top of the exhaust stroke to allow remaining gases to be drawn out by the vacuum created by the exiting exhaust gases in the exhaust manifold.

The myth I hear often is that valves set too loose are better than valves set too tight, because a valve set too tight won't close. If the valve won't close, hot gases in the cylinder would rush by the open valve at high velocity causing erosion of the edge of the valve. Then the valve won't seat and compression goes down in the cylinder. This is true, but it doesn't justify the practice of setting valves purposely loose because the best performance would be achieved with the valves set as tight as possible without being too tight. I generally find that tuning an engine for best performance is highly correlated with a cooler running engine. In my experience, a tight, accurate setting at the manufacturer's recommended value is the best approach to valve setting. This still leaves quite a few thousandths of clearance as a margin for error.

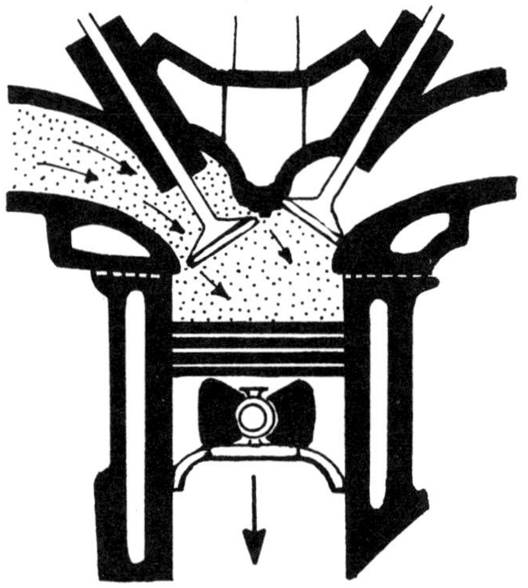

INTAKE

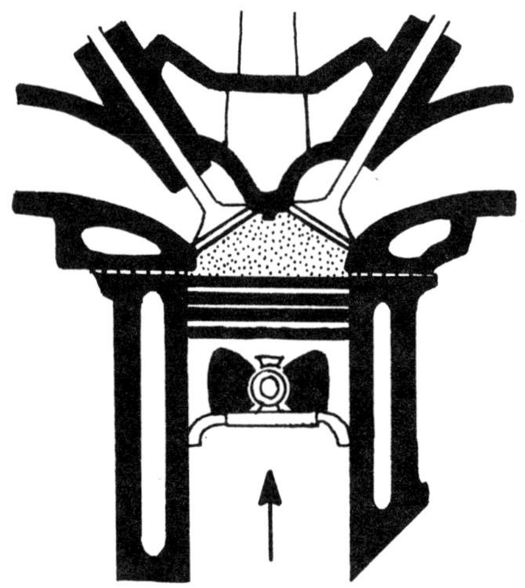

COMPRESSION (SPARK)

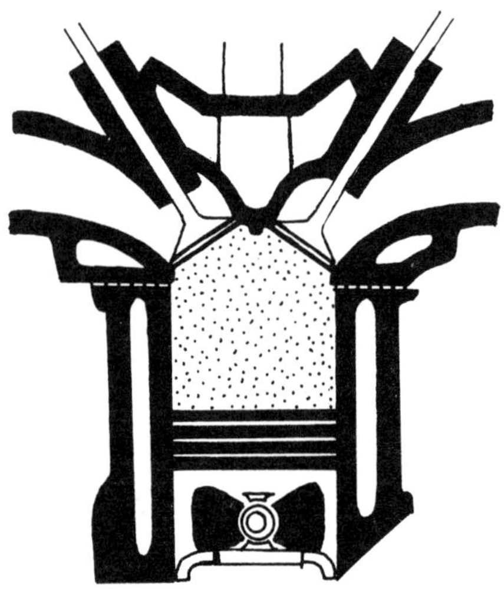

POWER

EXHAUST

I have discovered that a VW 1600, which was overheating because of poor air circulation and with the valves set normally, still had some valve clearance left; therefore setting the valves looser would have been superfluous. The real problem here was air circulation not adequate to handle heat load on the number-three cylinder. Finally I would like to quote A. L. Dyke, the publisher of the first practical book on automobiles in America, on this subject. "The value of the design of the cam, can and nearly always is, lost through improper valve clearance. . . . Many people who think, because an engine is new or has just been overhauled the timing must be right, will have a sad awakening if they will only spend a few minutes in verifying the timing."

A discussion of altering the shape of the cam to change the timing is beyond the scope of this book, but the other elements of timing—points setting, valve setting, and distributor timing—are basic to a good tune-up. If this is done accurately you will probably be bettering the standards of most dealer repair shops or weekend amateurs.

It's time to put the plugs back in, hook up the wires, put in the rotor, and crank up the engine. If it doesn't start, make sure you did put in the rotor, then check to see if the points are grounded. Remove the cap. With the ignition on and the points closed, attach the alligator clip of a test lamp onto an exposed part of the coil-to-distributor wire. Put the probe to the breaker plate of the distributor. The light should go on. Gently open the points with a pencil or small stick. The light should go off. The moving point should be electrically isolated from ground until it touches the stationary point. If the light doesn't go on in the first place there is an electrical break in the wires and connections somewhere. Lucas points are tricky in this regard. Be sure the condensor wire, the wire from the coil connector, and the metal spring of the moving point are in physical contact, but isolated from ground.

With the engine running, measure the dwell and correct the point gap if necessary. If the dwell is too high make the gap greater or vice versa.

Carburetor Adjustment

Take the air cleaner off and check the choke as the car warms up when it idles. The choke flap should be nearly closed initially, and within two minutes be *fully* open. Goose the accelerator linkage occasionally to allow the choke to open (for SU carburetors, see *Fuel Delivery II*). Whether hand operated or automatic, the function of the choke is to reduce the air portion of the gas-air mixture, and to increase the idle speed 200 r.p.m. or so above the normal warm idle speed.

All automatic chokes have a bimetal spring which is compressed when cold and expanded when hot. The differences between the chokes are in the manner in which they are heated. The VW has an electric heating coil connected from the ignition terminal of the coil. Chrysler puts the bimetal spring on the exhaust manifold and connects it to the choke by a rod. Other manufacturers use a metal air tube from the exhaust manifold to the choke housing. This tube conducts heat, but not exhaust fumes, to the choke. If exhaust fumes are coming through this tube, the choke flap may well be corroded into a fixed position that may require carburetor replacement. If the electrical unit is not heating, replace it. Check the connections and springiness of the bimetal spring.

The choke is adjusted by loosening three screws on the choke housing and rotating it until the choke flap is straight up and down when the engine is warm. The Chrysler type is adjusted by bending the connecting rod. Ideally, the choke should be on as little as possible.

When the engine is hot, adjust the air-fuel mixture screw. It's a unique screw made of brass, and is spring-loaded. Don't screw it in too tight. Any screw head which is fluted or knurled, as this one, is meant to be turned by fingers or, very gently with a screwdriver. If there is more than one barrel or throat in the carburetor, there may be two idle mixture screws (more in *Carburetor*).

With a tachometer hooked up to the distributor, screw in the mixture screw until the r.p.m. starts to drop, then screw it back out until you find a peak after which the speed drops again. Leave the setting on the peak r.p.m. coming to it from the lean side. This can be done by ear but not as accurately as with a tach. On some carburetors there is a definite peak, on others there is a plateau on the rich side.

Setting the lean-running emissions carburetors, which often have limiter caps on the idle-mixture screws, is a different story. Those plastic limiter caps are to keep people from setting the idle mixture in the traditional way as described above. The method for setting these carburetors is first to set the idle mixture for maximum idle with a tach (the dwell and timing having been set), then to adjust the idle screw to some speed above the final desired setting. This would generally be 40-50 r.p.m. for a carburetor with one mixture screws and 80-100 r.p.m. for a carburetor with two mixture screws. The correct idle speed is reached if you screw in the mixture adjustment (leaning the mixture) until the idle drops to the desired level. If the carburetor has two screws, turn them both in an equal amount. If you tried to set these carburetors in the traditional manner you wouldn't be able to get the idle low enough before the throttle butterfly would start sticking in the carburetor throat. For more information on lean carburetion consult *Smog Devices*.

Timing the Distributor

We are now ready to balance everything. Timing harmonizes all the various conditions, such as the type of gas used, temperature, altitude, engine age and so on. Ideally we would like the spark advanced as far as possible for any given condition just short of knocking.

"Knocking" is a detonation which occurs so early that it tends to push the crankshaft back around the other way. This is hard on the engine. If it persists, it will break a compression ring at least. Knocking sounds like cans rattling around and is diagnosed when the engine is under heavy load at low r.p.m., such as going up a stiff hill in second with the throttle wide open. Also, we want to avoid "dieseling", that is, running on after the key is turned off. A retarding of the spark sometimes solves dieseling, but probably only because retarding the spark lowers engine r.p.m. and therefore lowers the inertia of the flywheel and crankshaft. Dieseling can also be due to the timing set too retarded, which makes the engine run hotter.

This brings up a good rule of thumb about timing. When one is in the useful range of timing, namely when the engine will run, loosening the pinch clamp and rotating the distributor will cause the idle to increase in one direction and decrease in the other. The spark is being retarded in the direction that reduces the r.p.m., advanced in the other direction. Unfortunately, unlike the carburetor mixture setting, maximum r.p.m. achieved by twisting the distributor is not the best setting for the timing. It only runs acceptably at maximum r.p.m. because the engine is floating, so to speak. Under a load, it would knock.

There are two methods of timing an engine. Both require that you find the timing marks. Usually there will be a mark on a scale on the crank pulley, and a fixed pointer or ridge on the engine nearby. Sometimes one looks through a window in the bell-housing at a mark or scale on the flywheel. Most good repair books show a picture of what it looks like for a particular car.

With static timing you turn the crank by some means until the stationary pointer and the mark on the pulley line up, and the rotor is pointing to the number-one cylinder wire. Hook a test lamp onto an exposed part of the coil-to-distributor wire and stick the probe into some convenient ground. Loosen the pinch clamp and turn the distributor back and forth until the light goes out. Then turn the distributor slowly back to the point where the light just goes on. Tighten the pinch clamp, and the distributor is timed. A fair number of European cars use this method of timing. After an overhauling of an engine it is the method to use before starting the engine.

With dynamic timing you have to find the timing marks once again. If they are not very clear, they should be marked with chalk. Some timing marks are nearly impossible to see even at that. Next, hook up the timing light to the number-one spark-plug wire (preferably using a strobe light with an induction clip, battery operated). Loosen the distributor pinch clamp to the point where it will rotate but will stay in the position to which it has been rotated. Special wrenches are available to get at the pinch-clamp bolts on some American V-8s, but a crowsfoot will do.

Most American cars are timed with the vacuum line disconnected and plugged. I have a chain holding various-size plugs put together by Thextron. To start with, time the car to the manufacturer's recommendations. This will be adequate for most cases. Nowadays you have to look around inside the engine compartment for stickers indicating how to set the timing and final idle speed. Use the information on those stickers first. With the engine idling, the timing light hooked up, the distributor movable, and the timing marks identified, aim the strobe light at the stationary pointer on the timing cover. Rotate the distributor until the correct mark on the crank pulley is lined up with the stationary pointer. Tighten the distributor, recheck, then remove the timing light and reconnect the vacuum advance if necessary.

I like the timing method suggested for the TR-3 Triumph. The Triumph manufacturers suggest that you time the distributor statically with a test lamp according to the marks which show top dead center; then advance the spark by means of a calibrated adjuster on the distributor two notches (or 4 degrees); finally complete the timing adjustments by road-testing. Unfortunately only Lucas and some Japanese distributors have this handy little calibrated adjuster. In the old days there was a lever in the center of the steering wheel, called the spark, which allowed one to change the timing while the car was driven. One started out with the spark on or near top dead center, then one advanced the spark at higher engine speed. The spark has to be advanced from the setting made at idle speed because the rate of the explosion in the cylinder remains relatively constant while the piston speed increases with higher r.p.m. Therefore the spark has to occur earlier or more advanced of T.D.C. Full advance may be as much as 35 degrees before T.D.C.

This change in the timing with r.p.m. is accomplished by the centrifugal or vacuum advance mechanisms. You can check the action of these by holding the strobe light on the timing marks and, with the vacuum lines hooked up, increasing the engine r.p.m. The timing marks should change. The Mercedes-Benz engineers actually have you set the timing at idle as a first approximation, then again at 4,500 r.p.m. This insures that you get maximum advance at high speed.

The Peugeot engineers don't want anybody to experiment with the timing of their automobile. One rotates the flywheel until a metal rod, which you fit through a hole in the bellhousing, engages a notch in the flywheel. This is where a test lamp should just be turning on as indicated for static timing. I hope there is only one way to put on the flywheel. Older Chevy Sixes were timed through an inspection hole in the bellhousing by means of a ball embedded in the flywheel. Half the time the mark is now nowhere to be found, indicating that the distributor was probably put in wrong when it was rebuilt in the past. Since you can do the timing on any cylinder (the distributor-cam timing pattern is almost always the same for all cylinders) try clipping the strobe-light spark-plug lead on different wires until the timing marks appear in the strobe light. Then complete the timing as normal. Bosch distributors are occasionally installed 180 degrees out of phase by mistake if the engine has been rebuilt. It just means that the wires come out of the distributor cap opposite the position in which they normally are.

Some vans have timing marks in a place impossible to see. I time these by rotating the distributor to get maximum idle, then retard the timing around 200 r.p.m. Then I test drive. I call this "timing by experiment." If the timing is too retarded you will get hesitant or sluggish acceleration. Keep advancing the spark a little at a time until you get knocking. Then retard the timing a bit and you've got it. Some cars don't run worth a damn when you set them according to the specifications available to you, so in these cases you have to time by experiment.

Timing between the crankshaft and cam will be discussed in *Engine Overhaul*, and other timing strangeness in *Smog Devices*.

When the timing is completed, the engine warm and running smoothly, the final operation is to set the idle speed on the carburetor. The first approximation should be by using the manufacturer's recommendation. In the absence of that, set it so that the engine doesn't stall when you goose the accelerator and let it snap back, or so that the red ignition light goes out at idle on the older generator-charged automobiles. Ideally you want to have the engine idle as slowly as possible.

The idle speed is that amount of activity necessary for the engine to overcome its own internal resistance and to run its life-support systems, the oil, water, and fuel pumps, fan, generator and distributor. Increasing the idle is a good way to cover up engine weaknesses.

Older engines idled at much lower speeds than newer ones, the newer ones being required to operate smog pumps and lord knows what all. If an engine has to idle twice as fast to run its smog equipment, I wonder what actual over-all improvement has been made.

If the engine doesn't idle smoothly, if the carburetor adjustments didn't help, or if acceleration is hesitant, then it's probably time to rebuild the carburetor. A poor idle may also be due to corroded battery cables. Reaming and cleaning the battery clamps and posts could well be a regular part of service for a new customer.

Tune-up Check List

Basic minimum

- Check the spark plugs
- Check the compression
- Check the distributor cap/rotor
- Check the points/condenser
- Set the points
- Set the valves (solid lifters)
- Set the timing
- Set the carburetor
- Check the fuel pressure (fuel injection)

Additional

- Check the battery/starter motor
- Check the regulator/generator (alternator)
- Check the spark-plug cable resistance
- Ream the battery cable clamp and posts

Chapter 4:
Fuel Delivery

Never tamper with the carburetor until you are quite sure that it is at fault.
—Dyke's Automobile Encyclopedia

Fuel Pump

If you have a standard carburetor with an accelerator pump such as a Holly, Carter, or Solex, in fact any type except the SU carburetor or fuel injection, you can virtually eliminate the carburetor and the gas-delivery system as a cause for no starting or hard starting, by peering down the throat of the carburetor (air cleaner off), goosing the throttle linkage, and seeing if a squirt of gas is injected into the throat of the carburetor. Most no-starting or hard-starting problems are not the fault of the carburetor or the systems which supply gas to it. Even if you don't get a squirt of gas, then half the time it may simply mean you have run out of gas. Otherwise read on.

To see if gas is indeed getting to the carburetor, disconnect the gas line from the carburetor and turn the engine over a few times with the starter switch. Be prepared to catch the spurt of gas coming from the line with a bottle or a can. Generally a good healthy spurt with every turn of the engine is all that is necessary to give the fuel pump, fuel lines, and gas tank a clean bill of health. But if there is some question whether the volume is enough, because of apparent gas starvation at freeway speeds, you can either measure the volume pumped for sometime or measure the pressure generated by the fuel pump. This is done by inserting a pressure gauge into the gas line coming from the pump, then cranking the engine. It should be somewhere in the range of 4-7 lbs. Look in *Chilton's* for the particular manufacturer's recommendations.

Some fuel pumps used to have priming levers which allowed you manually to operate the fuel pump to see if it was working or to get fuel up to the carburetor after having run out of gas. Most modern American cars don't have these levers.

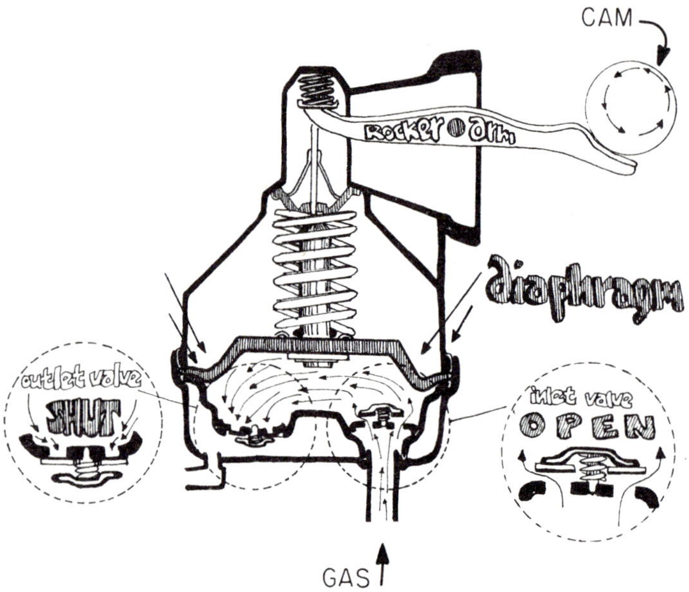

These days most fuel pumps are sealed units and are replaced as a whole. If the pump has a circle of screws around the outside edge, it can be rebuilt. Whether you can get the kit for it is another story. Call around first before wasting too much time on it.

The old types frequently had a sediment bowl as part of the unit. This allowed you to see if water was in the gas; if so it appeared as a clear fluid at the bottom of the bowl.

If you see water in the filter trap or the float bowl of the carburetor you can assume it will be in the gas tank. The fuel tank must be drained by removing the plug in the bottom; just drain out the first gallon or two since the water and other sediment is on the bottom. If somebody put sugar or some other soluble substance in your gas, then the whole gas system must be drained and cleaned. If poor performance is still noted then rebuild the carburetor.

Since fuel bowls are no longer a part of fuel systems you might buy a glass gas filter, available most anywhere, which has a cleanable reusable element. I use one, and it goes with me to my next car—along with my special shift knob, when I sell the car.

The fuel pump on those Chrysler slant sixes is a bear to get at. It is almost easier to loosen the engine mounts and jack up the engine a bit. I have seen the mounting bolts stripped on the fuel pump of this car. It can be done with the engine in place, so persevere. Some older Fords and American Motors cars had a double-action pump that created pressure on one side of the pump for gas delivery, and vacuum on the other side. The vacuum side of the pump assisted the vacuum-operated windshield wiper. Clever, what! This, however, led to oil fumes being sucked from the crankcase into the intake manifold if the diaphragm leaked. A smoky exhaust resulted, not because of bad rings but because of a bad fuel pump. This problem is peculiar to the Ford-type double-action pump. Nowadays we purposely vent the crankcase into the carburetor for smog control.

A more insidious problem with the fuel pump is a small hole in the diaphragm, not large enough to put it out of action, but enough to allow gas to leak into the oil sump. This leak thins the oil down and will cause severe ring wear in the few hundred to a thousand miles. You will smell the gas on the dipstick, but it will usually be too late; clouds of smoke from the tailpipe will tell you the engine has to be overhauled.

When you install a mechanical fuel pump, compare it with the old one. The activator arm which sticks into the crankcase leans against the camshaft. A node on the cam pushes it once every revolution. The activator arm on the new pump should be the same length and have the same curve as the old if you don't want to break the cam. When the pump is installed, it should abut the cam, not hook under it. If you want to be extra careful, turn the engine over by hand a few revolutions after installation to see that the engine doesn't jam.

Electronic Fuel Pump

Electrical fuel pumps are a different kettle of fish. They usually provide greater efficiency than the mechanical pumps, which makes them useful if you put multiple carburetion on the engine. Also, they may be a useful replacement when a mechanical fuel pump isn't available. The electrical pump should be mounted as low as possible without exposing it to ground hazards and as close to the gas tank as possible. The reason for this is that it pushes fuel forward better than it sucks it from behind.

The electric fuel pump begins to operate as soon as the ignition is turned on. You should hear it ticking away for a couple of seconds then die down to an occasional tick. If you have an accident, the fuel pump will keep pumping fuel, possibly through a broken gas line, as long as the ignition is on. The Vega electric fuel pump, which is inside the gas tank, doesn't turn on until oil pressure goes up. Current to the pump is routed through an oil pressure on-off switch like the brake-light pressure switch. The switch is screwed into the left side of the crankcase. If you have gas-delivery problems with the Vega, check to see if the plug is plugged in. In an accident, the engine will probably stop, oil pressure will fall, which will turn off the fuel pump. We may see more of this Vega-type fuel-delivery system in other cars.

Gas Tank

The fuel tank is usually no problem. It has to be vented so that the gas being removed is replaced by air. If the vent is clogged, it will be difficult to draw gas from the tank and it might even collapse. A gas tank may develop leaks. If so, it must be removed and replaced by a new one or one from a junk yard. No sealer I know of, including silicone sealer, resists the solving action of gasoline, with one exception: barge cement, which is used by shoemakers. Barge cementing a piece of woven fiber glass, such as is used in body repair, over the damaged area of the tank will probably work. Prepare the damaged area so it is clean and dry.

When I am restoring a car, especially one which has been sitting around for some time, I always put a gas filter in the line or replace the existing one, and I drain the gas tank. You never know what junk might be in the gas tank. Gas actually attracts water, and water raises havoc with carburetors. Some kind of chemical reaction goes on when water contracts the pot metal of the carburetors. The water creates a white passage-blocking residue which may be impossible to clean out adequately. You can put some methyl alcohol in the gas tank to dehydrate it.

The first engine I rebuilt had to be done over again because of some rusty junk in the gas tank which gummed up the engine. This was one of those crucial turning points for me. If one is not willing to bite the bullet and start over on a job even two or three times until it's right, then it's time to get a job with the post office. One often learns more from a failure than from a success.

If gas is gettng to the carburetor but not getting to the engine, it's time to turn to the next section.

Carburetors

I enjoy carburetor work. It is at the opposite end of the spectrum from clutch jobs, being lightweight, clean, and skilled work. The carburetor is actually a collection of subsystems whose function reveals much about the engine's varying need for fuel. If we understand how these subsystems work and when they fail to work, it will be easy to fix a carburetor, and it will be easier to understand and repair fuel injection systems. At the time of this writing carburetors constitute 80%-90% of the fuel-delivery systems on the road, but within a decade they may be rare as fuel injection takes over.

For self-teaching, the first carburetor to start is a simple single-barrel type as found on many American straight sixes or on the VW. Don't be disappointed with a 50%-success rate on your first few carburetors.

I usually do a tune-up first, which normally includes adjustment of the idle mixture and idle speed of the carburetor before I look to the carburetor as the cause of a problem. Ignition problems are much more common. If the car stumbles on acceleration, try advancing the timing a bit. If that fails to cure a ragged idle, hesitation on rapid acceleration, generally rough running, black exhaust; or if the carburetor adjustments didn't seem to help then you probably should rebuild the unit. Carburetor cleaners which you spray down the throat are a waste of time and money. The cleaner that you put into the gas tank may improve the carburetor's performance but will be hard on the neoprene parts and the gaskets. If you decide to rebuild the carburetor you should replace the gas filter at the same time. Often the opening of the gas lines dislodges dirt, and this dirt may foul even a new carburetor. Some carburetor manufacturers now insist that a new filter be used if their guarantee is to be in effect.

Carburetor Removal and Disassembly

To remove a carburetor is becoming increasingly complicated as additional smog stuff is attached to it. The first step is to disconnect the hoses going to the air cleaner. While you're removing the air cleaner check the cleaning element. If it is very dirty it might be choking off the air supply to the carburetor, making it run rich or, in the extreme, not at all. If there is a question in your mind just start the car with the filter removed to see if the problem disappears. Engines have been rebuilt which only needed a clean air filter.

Fortunately the various things hooked to the carburetor and air cleaner are usually unique in placement, size, or manner of connection so that they will remind you where they go. If you think there will be difficulty, mark things with tape. When you remove a solid gas line, as opposed to a rubber line with clamps, use two open-end wrenches. There are two nuts; one which has to be held, the other, smaller, one to be turned. If you just turn the smaller one you risk twisting the copper or steel line, causing an annoying gas leak. If the fitting is tight, use a flare-nut wrench such as used on brake lines.

Until you are adept at recognizing carburetor numbers, take the carburetor with you to the parts store or at least the small metal tag attached under one of the float-bowl screws. Kits are available for nearly all carburetors. They usually include all new gaskets, a new needle-valve assembly, an accelerator-pump diaphragm, ball bearings, and clips. The kits are designed to service several closely related carburetors hence include extra parts that will not be used. Do not look for new idle screws, throttle shafts, or any of the other substantial parts of the carburetor in these kits. If you need these parts go to the junkyard or visit a friend in the dealer shop. The dealers have their own private supplies from replaced carburetors.

Before you take the carburetor apart, see if you can't find a blowapart diagram of the carburetor on which you are working. For most American cars such a diagram comes with the rebuild kit. For foreign cars, a blowapart diagram will be found in *Chilton's* foreign-car reference book. These diagrams are useful to show where the ball bearing is supposed to go, which fell out when you separated the carburetor in two parts. Next, clean the dirt off the outside of the carburetor. This is the scrub-in before surgery. Carefully start dismantling the carburetor, putting the parts in groups in the order in which they come off. If you know their function you will be helped.

Float Chamber

The fuel pump draws gas from the gas tank and pushes it up to the carburetor. The fuel pump is designed to push more gas up than the engine can use. The carburetor has a small reservoir to hold the gas, called the float chamber, because of a float in it which rises to the top of the chamber with the incoming gas closing the needle valve. The needle valve controls the amount of gas that goes into the carburetor.

This method of metering the gas that goes into the carburetor is identical, as far as I know, for all down-draft or side-draft carbs (SU). The needle-valve unit is almost always a part of any carb rebuilding kit even though it is a rather reliable part because gas rushes by all the time, making the unit self-cleaning. To test whether it is leaking, remove the unit and blow through it in the direction of the gas flow and gently push in the needle part of the valve. It should close off the flow of air. Problems with the needle valve might be that the unit is loose or that the gasket is missing, allowing gas to by-pass the valve, overfilling the float chamber and flooding the engine.

On Holly carburetors several gaskets are associated with the needle-valve assembly. Inspect the blowapart diagram carefully. Don't assume that all the parts are in the correct place in your carburetor. The float is supposed to float and it usually does. If it is full of liquid don't bother to try getting new floats. Just buy another carburetor at the junkyard and use it for parts. The float setting is generally more of a problem if the carburetor has been "rebuilt" before. Evidences that the float is set to turn off gas flow too soon are gas starvation at high speeds and missing on a steep grade or on a hard corner.

I had an old Ford panel truck once which stalled on steep down-hill grades unless I lightly pumped the brake pedal, thereby sloshing gas into the appropriate jets, I assume. This occurred with a store-bought "rebuilt" carburetor. When I purchased this unit I could see light between the float-bowl cover and the float bowl—parts which are supposed to be sealed against gas leakage. This is a lesson I have come to learn over and over: just because you go into a store and pay your money does not necessarily mean that you get the solution to your problem.

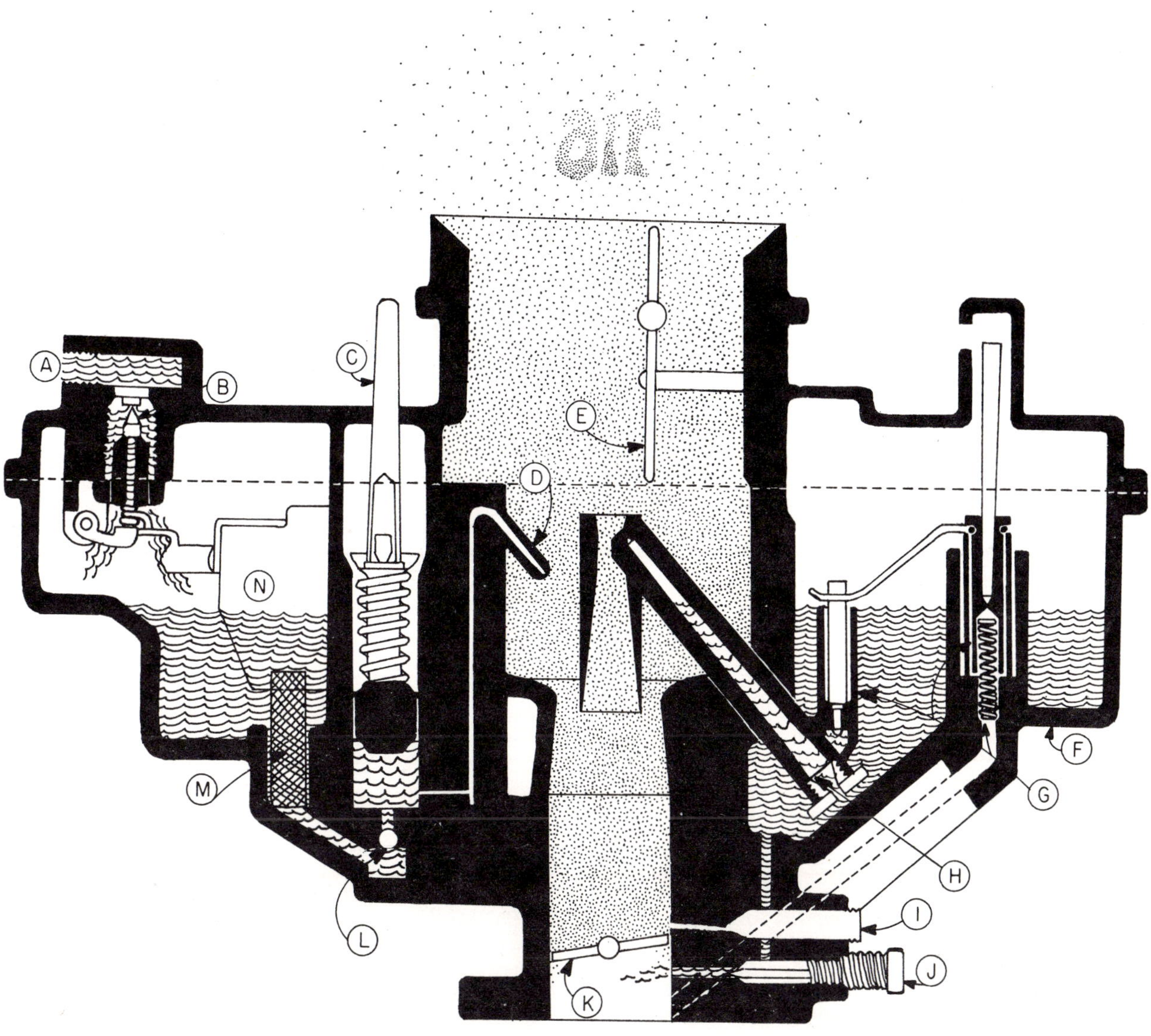

In particular, store-bought "rebuilt" carburetors are often poorly adjusted.

A float set too high doesn't turn off the gas. It gets pumped by the fuel pump straight through the jets into the throat of the carb, flooding the engine. Gas may also pour out the float-bowl vent tube. Look for and smell gas. A few carburetors have a glass float bowl or a glass window which allow you to check the float setting with the carburetor mounted and the engine running. Otherwise you check with a gauge included in the rebuild kit following instructions. Adjustments are done by bending metal tabs—crude but effective. Sometimes the amount the float drops is also controlled and measured in a similar fashion. Solex carbs, such as used on VWs don't have any arrangement for float adjustment. I've never seen it as a problem, which further suggestions that most float-setting problems on conventional carburetors may cur when they are fiddled with. When one has a double float arrangement, as with four-barrel carburetors, the floats should hang evenly. The arms may be carefully bent to achieve this.

Examine the bottom of the float bowl, which is a sediment collector. The amount and type of junk in the bottom will tell you whether you need a new gas filter and whether to anticipate dirt or water in the passages of the body of the carburetor.

Idle Circuit

A car driven in the city uses its idle circuit great deal even when it is moving. It is therefore important to have the idle circuit set properly—it can make a noticeable difference in gas mileage. The circuit has an adjustable mixture jet, which is a brass screw near the base of the carb with a knurled head and a spring. The knurled head means that it is to be turned by hand or by screwdriver very gently. When you remove it for cleaning, see whether there is a groove near the tapered tip. A groove would indicate that the idle circuit was screwed in too tightly, and the spray pattern will be adversely affected. I'm afraid you won't be able to get another except from a junk carburetor. The spring's function is to hold the setting once it is made.

There is another jet upstream of the mixture adjustor, called the idle jet; it is a brass screw with a calibrated hole, which meters the amount of gas needed. This jet can sometimes be changed for a bigger or smaller one. I will not discuss modifications here.

Solex conveniently use interchangeable jets in their various carburetors. If you lose the idle jet in your Porsche carburetor, the local VW dealer may have an exact replacement.

Throttle Butterfly and Shaft

The drilling for the idle jet is a small hole just below the spot where the throttle butterfly rests at idle. The throttle butterfly is connected by linkage to the accelerator pedal. At idle it is nearly closed. The extent of the opening is determined by the idle adjustment screw. The gap allows the appropriate volume of air to mix with the gas bleeding from the idle jet. Vacuum created by the pistons and a closed butterfly activate the idle jet.

On some newer carburetors the butterfly closes all the way to idle, and air goes through a bypass channel whose area is controlled by a bypass screw. A reduction of the area reduces the speed. The idle-mixture screw is situated in this bypass as well.

The throttle butterfly is mounted on a shaft which fits in two drillings at the base of the carburetor. The shaft is brass, and it wears after years of use allowing air to leak by. On high-mileage cars, check the shaft for play. Too much leakage will result in a rough idle. On occasion the throttle butterfly will come loose on the shaft, which will prevent the throttle from closing down properly.

Vacuum Ports

Below the idle jet there may be a drilling for a vacuum takeoff; this drilling is commonly on the intake manifold. This vacuum takeoff operated the vacuum windshield wipers in older cars. This port is connected to the tap to which you attach the vacuum gauge if you measure the health of the engine with the instrument. Vacuum in this area drops to near zero when the throttle is wide open. At idle the vacuum reading is normally from 17 to 22 inches.

Just above the idle jet may be one or two other drillings whose function is to bleed extra gas into the carburetor as the throttle butterfly opens. This smoothes the transition between the idle circuit jet and the main jet. If these transition jets are blocked, there may be a flat spot in performance at low engine speed.

Also located above the idle jet is the spark advance port. This vacuum drilling is connected to the advance diaphragm on the distributor. The vacuum here is low or zero at idle and increases as the throttle is opened.

Main Jet

The main jet meters the flow of gas from the float bowl to the nozzles that spray the gas in the throat of the carburetor. It is placed somewhere in the bottom of the float bowl. The main jet is often removable from outside the carburetor (in most Solex and Carter carburetors). The gas goes to the idle jet may pass through this jet but its calibration is for full-throttle operation.

Movement of gas from the float bowl to the main nozzle is caused by a vacuum created by air rushing through a construction in the throat of the carburetor called the venturi. The float bowl is vented to the atmophere, so atmospheric pressure pushes on the fuel. The more the throttle is opened, the greater the air rush, the greater the vacuum in the venturi, and the more gas will be pushed into the main nozzle. To facilitate the atomization of the gas as it passes through the drillings to the main nozzle, the gas goes through an air correction jet. This jet meters an appropriate amount of air to enter and mix with the gas stream. It would cause problems if you leave out these jets inadvertently when you rebuild the carburetor. If the main jet is left out or is loose, the carburetor will flood badly at any throttle position except when wide open. A flat spot in performance, as opposed to missing at upper engine speeds, may be due to lean main jets. Newer engines are being jetted leaner to get more complete combustion.

The main jet may have a calibrated needle that goes in and out changing the aperture of the hole. This is controlled by a little vacuum plunger and a fine spring. When rebuilding check to see if the plunger is free to move and the calibrated needles are straight. We have to be careful when you put the top and middle sections of the carburetor together, that the calibrated needles are going down into the main jet holes. If you are interested in modifying a carburetor, *Basic Carburetion and Fuel Systems,* may be of assistance. It is weak however, on the basic concepts of the functioning of carburetors.

Accelerator Pump

So far we described two steady-state systems. If we wished to change speeds with only the two systems so far described, we would have the cart before the horse. If we suddenly open the throttle, air will rush into the carburetor but there will be a short lag before the gas can be pushed from the float bowl into the main nozzle. Therefore, the cylinders will receive an extremely lean charge of fuel. In fact, the engine will die. This is where the accelerator pump comes in. This pump doesn't rely on vacuum for its operation. It is mechanically activated and operates when the accelerator pedal is depressed. Malfunction of this pump, causing hesitation on acceleration, is one of the most common symptoms of carburetor failure. This is because the pump has a diaphragm or a plunger, which can develop leaks. The leaking of the diaphragm is another instance where the mechanical portion of a system is the most likely to fail.

The accelerator-pump jet may be simply a drilling in the wall of the carburetor just above the venturi or it may be a tube that overhangs the opening and is aimed to hit the throttle butterfly in a certain way. Such is the case with the VW Solex carburetors. Here the jet is a press fit. It comes out and falls into the engine on occasion. In fact, I rebuilt my Karmann-Ghia engine once thinking something had happened to a cylinder. It turned out to be the accelerator pump jet which had rattled around in cylinder number three, then passed out or vaporized. As a mechanic I was able to remove and disassemble the engine. Otherwise it would have been one of those terrible engine worries that miraculously heal themselves leaving only a minor carburetor problem. New accelerator-pump jets are available for Solex carburetors at the dealer.

The accelerator-pump stroke can be adjusted on the outside in many carburetors. In the old days you changed your pump stroke with the season on some cars. This could temporarily solve the problem of not getting enough gas at the beginning of the stroke. When an American carburetor is rebuilt, the length of the accelerator-pump stroke is part of the normal adjustment procedure and is covered in the instruction sheet in the rebuild kit.

The accelerator pump primes your engine in the morning when you put your foot on the accelerator pedal. Hard starting may therefore indicate a bad accelerator pump. Remove the air cleaner and look into the throat of the carburetor. Open the choke flat and goose the accelerator linkage. You should see a strong steady stream of gas with the movement of the linkage.

The accelerator-pump subsystem includes the ballbearings which might have fallen out of the middle section as you separated the carburetor in two. They sit in the bottom of narrow wells or drillings on a seat which makes them one-way valves. As the pump pushes gas up through the accelerator-pump jet, one ball bearing gets forced against its seat keeping gas from also being cleared out of the accelerator jet. This system has its own opening to the fuel bowl apart from the main jet.

To cure a carburetor problem sometimes requires no more than replacing the diaphragm or the plunger. However, one has to be sure it works after it is supposedly fixed. On some carburetors the system will work when the middle section must be put together; on others the top section and the middle section must be put together. At any rate, I always put gas in the fuel bowl after cleaning and test the accelerator pump for a strong continuous stream before I completely reassemble it or before I install it in the car.

The ball bearings in the VW Solex carburetor are nasty to get at but if you've got water in the carburetor the chances are that the drillings, ball bearings, and jet are gummed up. When you look at the carburetor as it sits on the engine, there is a drilling along the lower right side of the bottom section (in a two-section carburetor there is no middle section) that goes straight forward ending in a plug facing you. At right angles going up to where the accelerator-pump jet sits, there is another drilling. At this intersection is one of the ball bearings. You can get at it by drilling out the plug. The jet can be removed by careful tapping and levering. Then the drillings, the ball bearing, and the jet can be cleaned. Buy another jet if the dirty one is intractable. Don't enlarge the jet hole trying to clean it—it is calibrated. After replacing the ball bearing, a new plug will have to be put in. I use a cut-off self-tapping screw with a fiber washer.

The other ball bearing sits under a brass plug in a shoulder near the bottom of the fuel bowl. It is threaded and will accept one of the accelerator-pump housing screws. Then you have to figure out how to pull in the screw without destroying the carburetor in the process. Fortunately, it is usually the other area which gives the trouble. If you mess up the carburetor doing all this, you have to get another one, anyway. Solex carburetors used on better cars like the Mercedes are more convenient because they use threaded plugs which are easily removed and replaced.

Choke

Now that we have the carburetor idling, cruising, and changing speed, we come to some of the accessories. The choke provides a temporarily richer mixture during cold starts. The flap in the top of the carburetor is the choke. It functions by closing off the incoming air supply. This changes the fuel air ratio to richer; so rich in fact that the car wouldn't idle properly if at all. Therefore a cam device is hooked to the choke which increases the idle speed, that is, opens the throttle flap a bit.

In a cold engine fuel doesn't vaporize readily, and even condenses on the walls of the cylinder. Hence the need for an enriched fuel mixture. The gas which doesn't burn in the explosion washes lubricating oil off the cylinder walls, passes the rings, and goes into the oil sump. Much engine-ring wear occurs at this time. The warmup period of most equipment, whether it is light bulbs, television sets, or internal combustion engines, is the time of greatest wear.

Ideally the machine shouldn't be driven until warm. The choke allows us to drive whether the machine likes it or not, but not without cost. It is important that the automatic choke is fully open when it is supposed to be, so as not to continue this enriched condition any longer than necessary. To adjust the choke it is necessary to loosen some screws and rotate a housing that contains a bimetal spring, or to change the length of a rod which connects the choke shaft to a bimetal spring. If you make the adjustment, pull back the throttle arm, so the choke shaft can move freely. The choke flap is fully open when it is straight up and down.

Even with the choke fully open the shaft cuts down on volumetric efficiency, that is the volume of air which can pass through the carburetor. To conserve gas, most average passenger cars are somewhat "strangled" in the carburetor-throat area.

The choke flap is closed when it is cracked open just a bit. Some air has to get in, hence the amount the choke should close often requires an adjustment detailed in the instructions that come with the rebuild kit. On older carburetors there was often a small spring-loaded flap in the choke flap which would open if you accelerated enough. This was for extra air. Some newer carburetors have a vacuum-operated choke unloader, which overrides the choke to a limited extent during hard acceleration. Other carburetors don't have a choke unloader. I believe it is necessary only for some designs of choke mechanism. Otherwise the choke flap, spring loaded as it is, tends to open by itself during the acceleration air rush.

If the choke-flap shaft is binding, it is important to free it. The screws which hold the flap in the shaft are brass and often staked so that it is difficult to remove them. If you are prepared to deal with drilling and tapping, the flap can be removed, but it would probably be cheaper to get a junk carburetor that has a freely moving choke flap.

On four-barrel carburetors the choke flaps are placed only on the two primary barrels or throats, and the secondary barrels, if progressively linked, are blocked from coming on until the choke is fully open. Some two barrel carburetors have a primary barrel that is choked, and a secondary barrel that is locked out until the choke is off.

When the choke is on the throttle, it is held open by a stepped or smooth fast idle cam, which increases the idle roughly 200 r.p.m. There is an adjustment screw to control this fast idle. Make sure you are off this cam when the engine is warm and the choke is off. On occasion I have ground off these cams carefully so that this would be the case. Try everything else twice, however, before you grind something off.

Dash Pot

The vacuum-operated carburetion system as compared with electronic fuel injection has this drawback: When you decelerate from high speed and use the engine for braking (operating against compression), gas gets sucked into the cylinders but, because of the closed throttle, goes unburned straight through the system. An unsatisfactory solution has been to put a vacuum-operated dash pot on the carburetor near the throttle arm, which keeps it from snapping down to idle when one's foot is removed from the accelerator pedal. This solution keeps engine revolutions high to burn up the gas, but eliminates engine braking. The electronic fuel-injection system cuts off fuel delivery at this time. I think a vacuum-operated switch along with an electromagnetic idle-jet cutoff switch such as used on VW Solex carburetors would be a better solution than a dash pot. One Japanese carburetor operates this way.

The dash pots on some carburetors may have worked in the laboratory, but in the real world they probably create more smog through malfunction and maladjustment than they were designed to eliminate.

Electromagnetic Cutoff

As idle speeds had to be increased to 1,000 r.p.m. to run smog equipment and air conditioning, the problem of dieseling increased. "Dieseling" means that the engine is running on after the ignition key is removed. It happens because increased engine inertia at the higher idle, coupled with the hotter engine (leaner main jets) and heat of compression, combine to cause ignition of the gas-air charge still being sucked in without need of spark at the spark plug. To eliminate dieseling, some carburetors have an electromagnetic idle jet which blocks gas flow to the idle jet when the ignition is turned off. To check this gizmo, turn on the ignition, find the cutoff solenoid on the carburetor, and pull off and put on the electrical connection. If you hear a click it is alright. This is my procedure for testing any electromagnetic switch or solenoid. Other indications of malfunction in this switch are if dieseling occurs or if the car will start, but not idle.

I have used an electromagnetic cutoff to eliminate dieseling on an older, high-mileage, abnormally-high-compression engine but not without experimenting with timing first.

Cleaning the Carburetor

Having cleaned off the outside of the carburetor with regular solvent, the next step is to soak the various metal parts of the disassembled carburetor in methylene chloride. I buy it by the gallon from a nearby chemical wholesaler, a small businessman.

I dump an appropriate amount into a large 5-pound coffee can, enough to cover the parts, then let them sit for a while. Avoid getting this stuff on your hands or in your eyes. It is a powerful varnish remover and oil remover. You can then spray all passageways and jets with an aerosol-spray carburetor cleaner or, much cheaper in the long run, get a Sure-Shot reusable spray can filled with methylene chloride (see *Tools of the Trade*). The stuff in the coffee can may be reused many times. After use, restore it to the original container, all except the last shot-glass full with the sediment. This you can throw away.

After cleaning, putting in new parts, and checking the action of the accelerator pump, screw in the idle jet until it is snug; then back it out 1½ turns. Now it is time to reinstall the carburetor. If the vacuum hoses seem a loose fit, cut off half an inch or replace them. I often start by hand a brass vacuum fitting while the carburetor is still loose, so I can get good alignment of the screw with the fitting. Brass fittings are so easy to strip. Don't overtighten. Adjustment of the carburetor is described in the tune-up section.

If the car starts but stalls almost immediately you perhaps left out a main jet or failed to put in or securely tighten the needle valve.

Two and Four Barrels

No matter how many barrels we have the same basic systems are present. The major difference between two- and four-barrel carburetors is how the barrels are coupled, that is, how they work together. Two-barrel carburetors have their respective throttle butterflies on the same shaft in the same body, hence coupling is no problem. Four-barrel carburetors are two two-barrel carburetors hooked together by a connecting rod. This rod is adjusted when the carburetor is rebuilt but has to be modified at times when performance requires it. Generally, on four-barrel carburetors the two sets of two barrels are hooked together progressively. For normal cruising only one set is in operation. When the accelerator is floored, the second set (called the secondaries) is brought into operation. The secondaries are identified by the lack of a choke butterfly and they are locked out of operation while the choke is on in the primaries. You can feel when the secondaries are cutting in (if they are), hence you can choose when they start by changing the length of the connecting rod. I will discuss other linkage, as a part of an over-all description of SU carburetors (which generally comes in pairs) in the following section.

SU Carburetors

The percentage of cars in the United States using SU carburetors are small, but they are found on some very interesting cars. Most British cars from the Bugeye Sprint to the Aston Martin and also the Volvo and Datsun 240Z use these carburetors. They are beautiful, simple, and reliable when they work but there is more art and ear needed to tune these than most other more common carburetors. Since they generally are seen in pairs and trios, the general approach to harmonizing multiple carburetors will be covered here. But first I want to relate their individual manner of functioning to the previous section.

Intake-Manifold Designs

The SU is a side-draft carburetor which means the air comes through the side rather than through the top. Ideally we would like air to flow into the cylinder as straight as possible for best performance. Every curve introduces turbulence, which resists flow. With the same piston suction less fuel will reach the cylinder with a curved intake manifold than with a straight one.

In the old Chevy 235 six-cylinder the top-draft carburetor sits on top of an intake manifold quite close to the head. The air rushes through the carburetor, makes a 90-degree turn at the intake manifold where it branches to the various cylinders, and then makes another 90-degree turn to go into the cylinders. That is a lot of resistance to flow. In addition, the length of the intake manifold between cylinders one and six and the carburetor is much greater than the distance to the two middle cylinders. This means that the outside cylinders will always be running lean compared with the middle two.

Now think of the Chrysler slant-six engine. Since in this engine, the carburetor is situated quite a distance from the head, the paths from the carburetor to the various cylinders are nearly the same. Also, the curves are more streamlined. However, the top-draft carburetor still sits on top of the manifold, introducing a 90-degree turn at that point.

Now picture the manifold even more straightened out, with a side-draft carburetor attached to the ends of the pipes, one carburetor for every two cylinders, and you have the typical SU setup. The air now takes almost a direct route into the cylinder. If the exhaust manifold were on the opposite side, more or less continuing the flow in a straight line, we would have a cross-flow head, which would be the most efficient. The design of the typical V-8 with a top-draft carburetor is only a fair approximation to a crossflow head.

SU Idle System

With the air rushing in the side of the SU carburetor, the venturi or construction is the same as in the standard top-draft type. The idle jet, however, is placed in the base of this construction. The vacuum created by the air rush sucks fuel directly from this jet rather than from a drilling behind the throttle butterfly. The fuel bowl and needle valve operate in the same fashion as any other, although they may not be part of the main body casting. A hose connects the bottom of the fuel bowl with the main jet. This hose is a source for leaks. A new hose and jet assembly come with the rebuild kit.

Coming from the top of the throat above the idle jet, a needle fits into the jet. At idle the jet's orifice is nearly filled by the needle. The needle is tapered so that the further it goes into the jet the smaller is the orifice through which gas may flow. The position of the jet in relation to the needle is adjusted by the jet adjusting nut which moves the jet up and down. In this manner the fuel/air mixture is set. Turning the adjusting nut up or tighter makes the mixture leaner; turning it down makes it richer. The idling speed is adjusted by a screw near the throttle linkage.

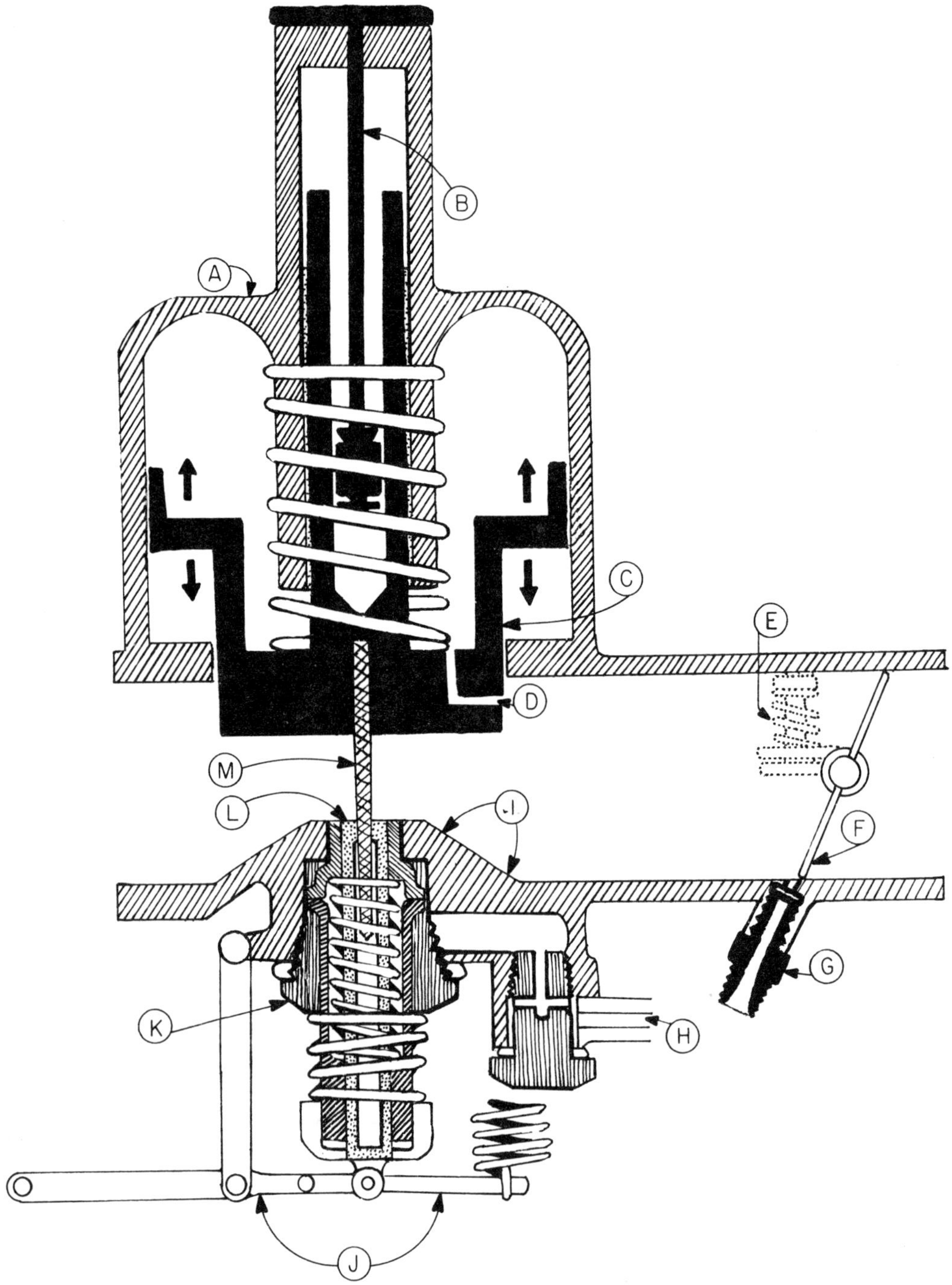

A Vacuum Chamber
B Piston Damper
C Vacuum Piston
D Vacuum
E Idle Adjustment
F Throttle Butterfly
G Distributor Spark Connection
H Gas Inlet
I Bridge
J Choke Linkage
K Mixture Adjustment
L Jet
M Tapered Needle

SU Main Jet

Actually the idle jet and the main jet are one and the same. What changes is the position of the tapered needle. (A Zenith carburetor, otherwise similar in design to the SU, uses a vacuum diaphragm rather than a vacuum piston.) The piston moves up and down depending on the vacuum in the chamber above it. The vacuum, created by air rush in the venturi construction, is vented through a drilling in the carburetor body to the vacuum chamber. The vacuum port is similar to the spark-advance port in the conventional carburetor in that the vacuum is near zero at idle, then increases with the throttle opening. The needle can be changed, thereby affecting richness of the mixture.

No accelerator-pump jet is needed on this design of carburetor because when the throttle butterfly is opened by the accelerator linkage, the vacuum piston lags momentarily; this lag forestalls the temporary leaning of the fuel mixture that occurs on the conventional carburetor. The amount of lag is determined by the viscosity of the oil in the piston damper on the top of the vacuum chamber.

SU Choke

Since this carburetor has no accelerator-pump jet, a pressing of the accelerator pedal does not produce a squirt of fuel into the throat. Therefore the choke is needed to start a cold engine. The choke is very simple. The choke linkage pulls the jet assembly down ¼" making the jet-tapered needle setting very rich. A pulling of the choke also moves the throttle to a fast idle position to keep the rich mixture from stalling the engine.

Trouble Shooting SU Carburetors

Three problems occur with the SU carburetor: First, a leak at the connection between the fuel bowl and the jet; second, a vacuum piston that doesn't move up and down freely; third, a choke mechanism that doesn't move up and down freely. In the last two problems the relationship between the needle and the jet is inappropriate to the engine demand for fuel. The mixture will generally be too rich.

As for the third, a stuck choke mechanism may wipe out new rings in 2,000 miles. Also, I had a Volvo in which the front member of a pair of SU's developed a hung-up piston. Fuel ran so rich in the front two cylinders that it condensed on the cylinder walls, went past the rings, and diluted the oil so that it bled by the oil-pan gasket, covering the under side of the car with a thin gasoline/oil mixture. Oil pressure dropped, of course, and rings on the front two cylinders wore badly in only a few hundred miles.

As part of a tune-up the airfilter should be removed and replaced if dirty. The pistons now exposed should be picked up with a finger and let drop. They should move up freely if somewhat stiffly and fall promptly with a slight "thunk" as they hit the bridge. The bridge holds the jet assembly in place. The choke mechanism should be pulled down and let snap back. It should snap back because it is spring loaded. The damper should be checked for oil. The act of checking and especially refilling it with 20-weight engine oil lubricates the piston rod. If this process reveals a discernible insufficiency, consider rebuilding the carburetors. If the spark plugs are running black, this is additional motivation to rebuild.

Rebuilding SU Carburetors

Rebuild one carburetor at a time. Don't interchange any parts. Save your most gentle touch and clean hands for this job. Mark the position of the vacuum chamber because it has to go back the same way. Sometimes the screw positions determine the position of the vacuum chamber. If you notice that the screws holding down the vacuum chamber are damaged by overtightening, look out! The chamber has been opened by an amateur who thought the screws had to be tight to maintain the vacuum in the chamber. They should not be. If the carburetor hasn't been opened since the factory put it on, you are in luck.

Clean everything with a powerful varnish-cutting solvent such as methylene chloride. If the piston is hanging up, look for dirt between the piston rod and the inside bore of the suction chamber. rather than for problems between the edge of the suction piston and the outside bore of the suction chamber. The reason for the piston rod hanging up on the inside bore is the closer tolerances observed at the piston rod as well as the presence of oil in the area. Clean oil is a lubricant, dirty oil an abrasive.

I strongly recommend the purchase of the rebuilding kits made by SU, although they cost four times as much as others. They include a new jet assembly, gaskets, a needle valve, and a needle. Be sure you know the details of engine capacity and the year of the car. The throttle shafts wear out on this carburetor as on any conventional type. If the idle below 1,000-1,200 r.p.m. is not smooth, the throttle shafts may be worn. These are supposed to be available but you may break your head trying to find them. If an air leak exists, you may compensate by setting the carburetor rich, which on the SU carburetor means it will be rich throughout its full range. Look for carbon black on the tailpipe or plugs.

I don't always replace the needles. The needles tend to be self-cleaning since gas rushes by it all the time. If, however, the jet hole appears to be ovalized, replace both the jet and the needle. In some of the older jet designs where the gas didn't come in through the bottom but from the side, the bottom of the jet would collect sediment, which could hang up the needle. If you wish to replace the needle, loosen the set screw and dispose of the old one. The untapered shank of the new needle may be a tight fit. If this is so, twist it around in a piece of 320 emery paper to reduce the size just a hair. Don't tamper with the tapered section.

Reinstall the jet assembly but don't tighten the jet locking nut. Be careful of the needle and carefully install the vacuum piston. Now replace the spring and vacuum chamber, lightly oiling the piston shaft. Spray the jet locking nut and the general area with WD-40. Screw up the mixture screw to the leanest position. With the jet up as far as it will go, the locking nut a little loose, move the piston up with a finger and let it go. Move the jet assembly a little this way and that until the piston falls freely hitting the bridge with a thunk. This may take a little doing, but don't continue until it works. If no amount of finagling works, try switching the vacuum chambers because they might have been reversed by a previous "rebuilder."

When the vacuum piston moves freely up and down, tighten the locking nut. Check it again. If it is still moving freely, fill the hollow piston rod with 20-weight oil or whatever is recommended. Put in and screw down the damping rod.

In a few instances the rebuilt jet assembly will fail to pull back up freely into its normal position when the choke linkage is slack. I have solved this by adding another long light spring to the assembly on the side opposite the existing spring. This spring mainly counterbalances the bias of the existing spring but hardly adds more spring pressure. The jet mechanism can get cocked when pulled up from only one side.

Emission SU Carburetors

The new SU carburetors cope with emission standards in the following way: A spring-loaded needle is incorporated, which makes the needle biased in the jet, that is, it is not centered. The jet itself is not laterally adjustable. Cold-start enrichment is not accomplished by lowering the jet. The new SU carburetors have now a separate enrichment device that simply opens the gas flow through another drilling and a separate jet from the main jet. The float chamber is now part of the main carburetor body casting. Mixture is controlled by a screw rather than a nut. A throttle bypass helps atomization of the gas/air at low throttle and an overrun throttle valve. This is a little spring-loaded valve in the throttle butterfly, which opens during rapid engine decleleration when there is a high intake-manifold vacuum. This valve lets in some air to help burn the temporarily rich mixture. Finally, a bimetallic spring in connection with the mixture setting screw adjusts the jet according to temperature. This adjustment compensates for changes in gasoline viscosity. The warmer the gas, the more readily it flows through the jet.

As a result of the changing environment and cost of gas, carburetors have had to become more accurate controlling devices. Performance alone can no longer be the criterion for a healthy carburetor. It is unfortunate that the engine runs best on a slightly rich mixture, which produces more harmful emissions. A slightly rich mixture gives a cooler, more controlled, burn as opposed to a hot, power-robbing detonation.

Tuning SU Carburetors

The first step to synchronize multiple carburetors is to disconnect the linkage connecting them. Then each carburetor is adjusted for idle mixture and speed separately. The engine, of course, has to be thoroughly warm. Since the mixture screw is all the way up after the rebuilding, start by screwing it down two full turns. The use of a tachometer makes this job easier. Set the idle-speed screws to achieve an idle of around 900 r.p.m.

To set the mixture, I screw the mixture control in or out until I find the maximum r.p.m.; then I screw the mixture control back in to the lean side, until the r.p.m. falls, then back out again until the maximum previously determined is just reached or even a quarter turn beyond that. When this is done with the other carburetor, I repeat the process with the original one. Going back and forth is necessary because there is some cross-influence caused by the interconnection in the intake manifold.

Now I reset the idle-speed screws, turning both an equal amount until 800 or 900 r.p.m. is achieved. I listen to the exhaust note—it should be smooth and even. Irregular, is too lean; rolling or surging, is too rich. I fiddle with the mixture screws until the best sound is achieved.

An equal idle speed on each carburetor is achieved by a two-foot length of heater hose. Listen to the sound of the air rush; one end of the hose to your ear, the other at the mouth of the carburetor. When the same sound is heard in each by the adjustment of the idle-speed screws, then the two carburetors are balanced. A Uni-Syn is a device that allows this to be done visually.

Road test the car and readjust as necessary. I find that I have to tinker with the idle speed after a few days of running. The ignition system must work properly before a good carburetor adjustment can be achieved.

Chapter 5:

Fuel Delivery II

When neither statistics nor probability theory existed, abracadabra could flourish uncontested.

—Leo Rosten (b. 1908)

Fuel Injection

Fuel injection of some sort seems to me to be a wave, or should I say a spray, of the future. The factors forcing the switch to fuel injection are that the fuel flow can be cut off completely on deceleration, and that fuel can potentially be more accurately metered over a wider range than is possible with conventional carburetion. The first factor cuts down on harmful emissions, the second gives better horse-power output per cubic inch displacement of the engine.

In practice, performance doesn't always improve. Some of the first VWs with electronic fuel injection got terrible gas mileage and performance right from the start. Many of these difficulties have been ironed out. There is probably another reason for switching to electronic fuel injection. The dealers made the system a mystery and are now in a position to corner the market on adjustment and repair. Fortunately, the demands of the engine remain the same, and a mechanic who understands the function of the conventional carburetor will have no difficulty with fuel injection.

First, let me emphasize that before the electronic injection system is worked on, a tune-up should be done to eliminate the much more common and potentially confounding ignition problems.

Fuel Pump and Regulator

Fuel pressure in the injection system is higher than in the normally carbureted engine. Besides, it plays a crucial part in determining the amount of gas delivered by the injectors. In the conventional system, the fuel pump merely gets gas to the float bowl. Only in the case of a damaged needle valve would fuel be directly forced through the jets as it is in the fuel-injection arrangement.

The fuel line leads directly to the injectors and thence to a fuel-pressure regulator that can bypass fuel back to the gas tank thereby maintaining the pressure at the injectors at a constant value. The rate at which the fuel is sprayed into the intake manifold depends on the time the injector is open and the pressure in the gas line. Increase the pressure and more fuel is delivered. Increasing the pressure is comparable to enlarging the main jet in a carburetor. When Saab went from a 1.7- to a 1.85-liter engine, nothing was changed in the injection system except to raise the fuel-pressure setting of the regulator from 28.5 to 32.7 pounds per square inch (p.s.i.).

The pressure can be checked by the insertion of a pressure gauge between the injectors and the pressure regulator. Simply turning on the ignition key activates the fuel pump for a moment. Turn the key on and off a few times to build up pressure. If the pressure is incorrect for the particular car, turn on the engine and idle, loosen the locknut on the pressure regulator, and turn the adjusting screw until the correct pressure is read on the gauge. The locknut is then retightened.

Turn off the ignition switch and see if the pressure drops below around 17 p.s.i. If it does, the gas line leaks somewhere, usually at a clamped joint. I recommend wholesale replacement of all flexible gas lines and clamps on an older car when leaks start to appear. Get an 8-foot length of gas line from the dealer to make sure you have the right size. Replace the short hose on the injector by cutting off the crimp fitting and restoring it with a short piece of conventional gas line and a clamp. Checking the high and low fuel pressure could be part of a regular tune-up.

When the fuel pump fails in the injection type of fuel-delivery system, the car cuts out abruptly the way it would if the ignition system failed on a car without fuel injection. In a conventional fuel system, fuel-pump failure feels like running out of gas: the motor sputters, catches, sputters, then dies.

The fuel pump may be in the gas tank. To test the pump directly refer ahead to the section *Pump Relay*.

Injectors

The injectors are in the intake manifold, and inject just behind the intake valves. They are electromagnetic switches. The injector plunger moves only .012", which should suggest how important it is to keep the gas clean by replacing the gas filter with every tune-up. If the spray pattern is greatly altered or blocked by dirt or water corrosion, the car will be difficult to start and perform roughly. The spray pattern can be directly observed with the injector removed and the ignition key turned on and off. Each turn on of the key will produce an injection.

Along with just about everything else in this system, the injectors are fixed by replacement. Bosch injectors are interchangeable between some different car manufacturers, but you have to pay attention to the Bosch part numbers.

Also, part of the fuel loop is an enrichment device that injects gas into the intake manifold for cold starts. Malfunctioning of this device produces symptoms similar to those of a bad choke on a carburetor.

Air System

Filtered air comes into a common duct which branches off to each individual cylinder. A butterfly valve is built into the intake duct which is connected by linkage to the accelerator pedal. The valve is like the throttle butterfly of a carburetor but it controls only air volume. Air and gas mixing occurs just before the intake valve. At idle the butterfly is completely closed and the idle-air volume is controlled by the idling-adjustment screw which is threaded in a

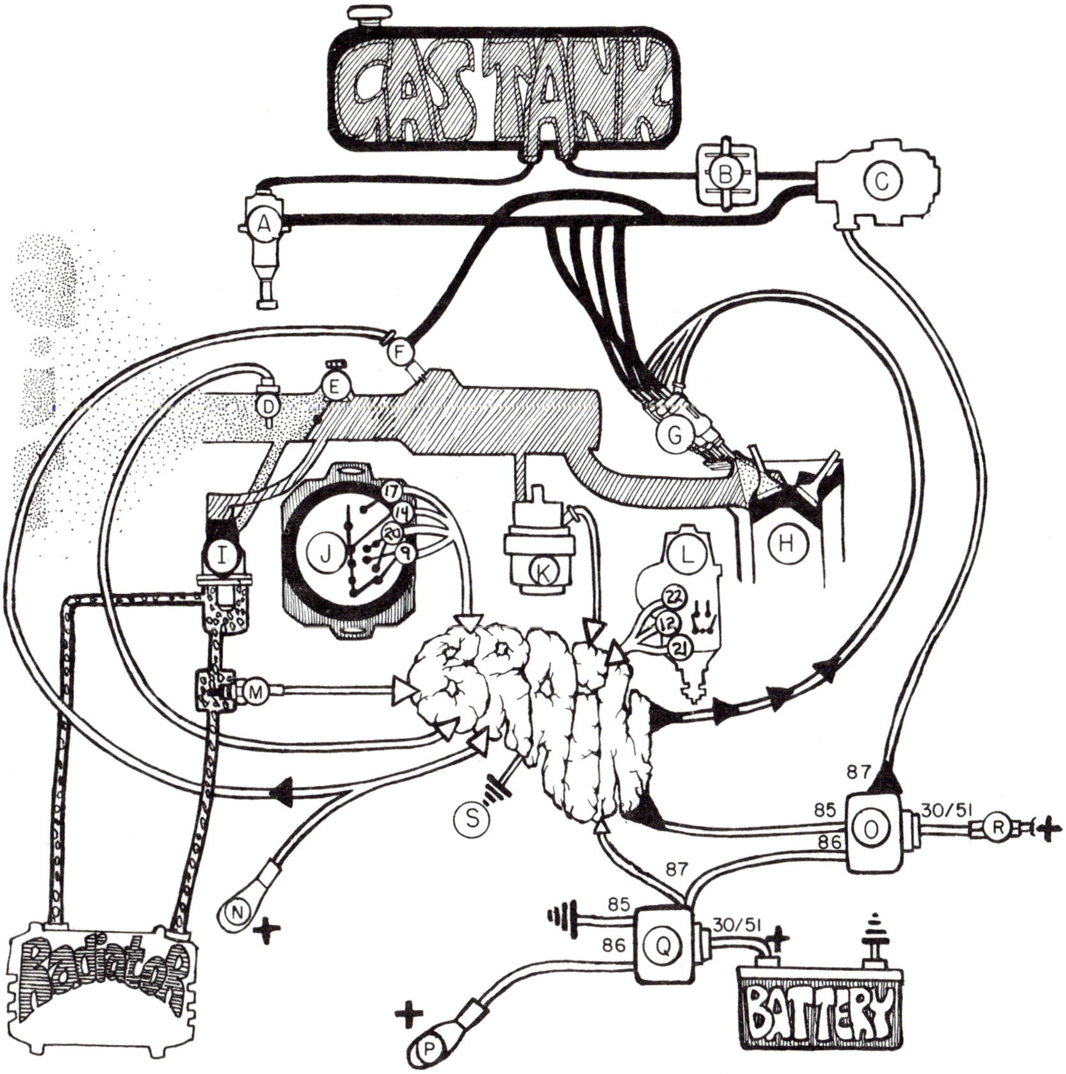

A Pressure Regulator

B Gas Filter

C Gas Pump

D Air Temperature (13)

E Idle Screw

F Start Valve

G Gas Injectors (3, 4, 5, 6)

H Cylinder

I Air Regulator

J Throttle Switch (17, 14, 20, 9)

K Pressure Sensor (15, 10, 8, 7)

L Distributor (22, 12, 21)

M Temperature Sensor II (23)

N Starter Terminal 50 (18)

O Pump Relay

P Coil Terminal 15 or Ignition Switch

Q Main Relay

R Fuse

S Ground (11, 26, 27, 28, 29, 32)

All numbers in parentheses are brain connector terminals.

little air bypass. This adjustment is usually all that is necessary for a regular tune-up.

Another air bypass is controlled by the auxiliary-air-regulator, which introduces more air during cold-start enrichment. It closes off as the temperature rises and enrichment ceases. It increases the idle speed and so functions somewhat like the fast-idle cam or a carburetor. A failure in this device would show itself in difficult idling when the engine is cold or in too fast idling with a warm engine. The normal position of this switch is closed. On water-cooled engines, water circulates through the switch to activate it. On air-cooled engines like the VW, a temperature sensor is bolted to the crankcase. This subsystem has a small diameter hose coming from the air cleaner to the switch and from the switch to the intake duct after the throttle valve.

To test this switch when cold, start the engine and remove the small hose from the intake manifold. If the idle improves, the switch is not opening as it should. To test it when it is hot and the engine seem to idle too high, remove the small hose from the intake manifold and block the opening with your finger. If the idle drops, the switch is not closing properly.

The Brain

The main function of the brain is to integrate information from the various sensing units to increase the number of injections or the time of their opening. It is reliable as long as it is not subjected to current surges, reversed polarity, or excessive heat. Always disconnect the negative battery strap when working on any part of the electrical system. It is usually a sensing unit, which is prone to breakdown, but if the brain malfunctions it is repaired by replacement. It is to be suspect only after you eliminate the sensors as described below. An ohmmeter check sequence, described at the end of this section, determines the health of the various sensors.

The current to run the brain comes from the main relay. Terminal 85 of the relay is grounded. Terminal 86 connects to the ignition switch or terminal 15 of the coil. Terminal 30/51 connects to the battery. Terminal 87 sends current to the brain box and the pump relay. Check with a test light whether this terminal is "hot," if the car won't start and the fuel pump isn't operating. Turn the ignition switch on first.

Pressure Sensor

This important sensor is found in the intake duct after the throttle valve, and it measures negative pressure, that is, vacuum. With the throttle closed, the vacuum is around 20 lbs., and with the throttle wide open, it drops to near zero. This sensor transmits information on engine load to the brain. Another function of this unit is to enrich the fuel mixture at full throttle. In the earliest injection systems this was accomplished by a separate pressure sensor, comparable in action to a power jet in the carburetor.

When this unit malfunctions it could result in the car failing to start, stalling after it does start, or failing to reach full performance. Depending on how it fails it could result in either flooding or starving the engine.

Trigger-Contact Assembly

This assembly, housed in the distributor, measures engine speed and triggers the injections. The injections occur in two pairs with a four-cylinder engine. Cylinders one and two receive injections at the same time, and cylinders three and four receive injections at the same time. The arrangement is done this way to simplify the electrical circuitry. If the contact resistance gets too great, the engine may start but stall, and will cut out when driven.

The basic duration of the injection is determined by the two previous sensors, which measure engine speed and load. Additional special conditions require additional sensing units.

Start Valve

This valve injects fuel into the common air-inlet duct for as long as the starter is operated and the temperature switch is closed indicating the engine is cold. In this way it is possible to get a cold start, then stall if either the trigger assembly or pressure sensor are malfunctioning. Cold-start difficulties may be due to malfunctioning in this valve. Some cars have a cold-start relay, which operates this valve rather than the brain box doing it directly.

Throttle-Valve Switch

This switch is attached to the end of the throttle-valve shaft. It informs the brain on the position of the throttle. It signals when the throttle is closed, and if this information is combined in the brain with the information from the distributor that engine speed is high, the fuel delivery will be switched off. This is the engine deceleration situation during which raw gas gets sucked unburned through a carbureted engine. When engine speed drops to 1,000 r.p.m., the fuel delivery is switched on again. This switch also serves the function that the accelerator pump serves on a carburetor. When the throttle is being opened, electrical pulses are sent to the brain, which delivers extra injections of fuel.

If this switch is malfunctioning or maladjusted, the results are misfiring on acceleration, backfiring on deceleration, or high fuel consumption. If this swich jams, it can cause the accelerator pedal to operate roughly or even to get stuck. The switch is supposed to begin operation just as the throttle valve has opened 2 degrees. To adjust it, first loosen the screws which hold it in position. Close it completely, then open until you hear a click; turn one more gradation on the scale, and retighten. Disconnect the throttle return spring and operate the throttle valve shaft to make sure it isn't binding. Reconnect the spring and you are finished.

Temperature Sensors

A car has two temperature sensors and one temperature switch. The switch was mentioned above (auxiliary-air-regulator). It controls the air bypass to increase idle speed on a cold engine. It is not controlled by the brain. The first temperature sensor measures the air temperature in the intake manifold, and below 5 degrees causes additional fuel injection by the start valve. The second temperature sensor measures coolant temperature in a water-cooled engine or oil temperature in an air-cooled engine. Until the engine is warm, this second sensor causes a lengthened injection period through the agency of the brain. The temperature sensors fulfill the role of the bimetalic spring in the choke on a carburetor.

Failure in these units could cause either a hard starting or a too-rich mixture on a warm engine. If the car has a rolling surge at idle it is probably due to a rich mixture.

Pump Relay

When the ignition key turns on the main relay, terminal 86 of the pump relay is made "hot." This can be checked with a test light. The relay doesn't close until terminal 85 goes to ground, which it does through the brain. The brain allows it to happen for 1-2 seconds after the ignition is switched on and whenever engine speed is above 200 r.p.m. These are the only conditions under which the fuel pump is supposed to operate. To see if the brain is doing its part, hook a test lamp between terminal 85 and 86. Turn the ignition on to see if you get a short flash on the test lamp. If you don't, check the actual grounding at the brain box. Double-check by grounding temporarily terminal 85 with an alligator test lead. If the fuel pump now works, replace the brain box. If the fuel pump still doesn't work read on.

Next, check terminal 30/51 (see previous figure). This should be hot with the ignition on. If it is not, follow the wire back to its source, which is a fuse. Check the fuse and its wiring. If terminal 30/51 is hot, attach a test lamp from terminal 87 to ground. Flick the ignition on and off. If the test lamp flashes, then the problem can only be in the fuel-pump connections or the motor.

You can bypass the pump relay by hooking the wire going to terminal 30/51 onto terminal 87 instead. However, the fuel pump will run all the time the ignition is on. This is dangerous in an accident, and inconvenient when a cylinder gets filled with gas through a defective injector.

Ohmmeter Check Sequence

If you have an ohmmeter or a Peerless model 465, very simple checks can be made on the sensing units to see if they are healthy. Remove the brain from its secret hiding place. Open the cable clamp with a Phillips screwdriver and slide off the plastic cover toward the cable. Carefully remove the cable-harness plug. The brain is now disconnected from the car. Put it aside.

You should have a wiring diagram similar to the next figure for the particular car being serviced. The diagram is for a Saab 1970, but the various types are similar.

First, turn on the ignition key. Test the plug connectors 16 and 24 with a test lamp or a voltmeter. A voltmeter should read 1.26 volts if the main relay, ignition switch, and battery are healthy.

Hook up the test lamp or voltmeter to plug connector 18. Crank the engine a few times with the ignition switch. The reading should be around 10 volts if the starting system is healthy.

With an ohmmeter connect one lead to ground and the other to plug connector 3, 4, 5, and 6 sequentially; they should each read around 2.4 ohms. These are the injectors.

Next, connect to plug connector 7 and 8 sequentially with one lead to ground. Both readings here should be infinite. If they are not, there is a short to ground. Now connect the two ohmmeter leads to bridge 8 to 10. This reading should be approximately 350 ohms. Now bridge plug connectors 7 and 15. The reading should be approximately 90 ohms. These readings test the pressure sensor.

Next, connect one lead of an ohmmeter or a continuity tester to plug connector 12 and the other to plug connector 21. Crank the engine a few times. The reading should alternate between zero and infinity. Change the lead on connector 21 to connector 22 and repeat the test. This measures the functioning of the distributor-trigger contacts.

Next, hook one lead to ground and the other to connector 23. This is the temperature sensor which in a water-cooled engine is in the cooling system somewhere. In the VW it measures head temperature. Its normal resistance changes with temperature, but it should be approximately 2,500 ohms at 68 degrees .

Next, hook to connector 13. This is the temperature sensor in the air-intake manifold. It should measure around 300 ohms at 68 degrees.

Next, hook a lead to the ohmmeter to connector 14. Hook the other lead first to connector 9. The reading should be zero and go to infinity when the throttle is opened. Then change the lead on 9 to connector 17. The reading should be infinite and go to zero at fully open throttle.

Finally, change the lead on 17 to connector 20. As the throttle is slowly opened, the ohmmeter should alternate between zero and infinity about 10 times. This test measures the functioning of the throttle-valve switch. This switch has all the functions of the mentioned newer fuel-injection systems. Originally all it did was tell the brain that the throttle was closed (the first test mentioned above).

If any of these checks gives readings other than those mentioned above, disconnect the plug on the unit involved and check it directly. If the reading is still wrong, replace the unit. If the reading is correct, then the wiring harness is at fault. The book says that any failure in the wiring harness means the whole harness has to be replaced. This is an enormous job for just one bad wire. The book claims that this is necessary because wire splices will introduce resistance into the system and it will not operate properly. I would say, this would not be a problem at all with the purely on/off type units such as the throttle-valve switch or the trigger contacts of the distributor. Neither would it seem to be important with the wiring going to the injectors. True, added resistance in the circuits from the temperature sensors or the pressure sensor will give the brain faulty information. Still, if you solder your connections carefully, checking performance and gas mileage after the repair, I would think it worth a try.

Intermittent Problems

The worst problems are those that occur only intermittently. If the problem is not manifesting itself while you test, there is no way to distinguish between the various components that may be responsible for the symptoms. Unless the problem exists at the time of testing, everything is guesswork.

The worst drawback of the use of an electronic brain, and therefore of the electronic fuel-injection system, is an intermittent problem in that unit. When it is replaced and sent to the factory for repair the brain may operate properly. The manufacturer will say that the mechanic made a faulty diagnosis and will return the unit to service checked and " rebuilt." You may get one of these units to replace a faulty one of your own. I know of a case where the dealer replaced an intermittently failing brain box by another intermittently failing brain box. I needn't tell you the anguish and inconvenience, not to mention expense, that this problem caused.

I called this the double-whammy situation. The first whammy is the intermittent problem. The second whammy is that after replacing a part of a system you tend to assume it is fixed, and look to all other possible causes of the problem before returning to the "new" part.

Continuous-Injection-System

Continuous-Injection System is the fuel injection now used on many European cars. We don't have enough experience with it to know how it will hold up in the long run, but from what I can surmise from the technical description, it seems to be one of the best innovations to come down the pike in some time. For one thing, it is rare that a new device is actually simpler than that which it replaces.

The fuel pump forces fuel under high pressure to the fuel distributor. The fuel distributor then feeds the four injectors which spray continuously as long as the engine is running and the pressure in the line high enough. Continuous injection is only feasible if the injectors can atomize the fuel to such an extent that it doesn't condense on the walls of the intake chamber. I believe that the high fuel-line pressure is necessary for this atomization to be accomplished.

To deal with the high pressure in the lines, special banjo fittings are used. The fiber gaskets should be replaced whenever the lines are removed or replaced. Another device is needed to compensate for high pressure. It is the fuel-pressure accumulator in the fuel line just after the fuel pump. Its function is to buffer the hammer blow of initial pressure (64 to 74 p.s.i.) that surges through the line when the ignition is turned on. It also maintains a pressure in the lines after the key is turned off to lessen the possibility of vapor lock.

The normal operating pressure is maintained at 48 to 54 p.s.i. by the pressure-regulating valve in conjuction with the pressure-relief valve in the fuel distributor, and by the pressure regulator (a separate unit) which changes pressure to compensate for temperature changes.

Pressure regulation is needed primarily to keep the pressure at the metering parts, and therefore at the injectors, constant. When the size of the ports is enlarged as the control plunger uncovers them, more fuel is allowed to pass to the injectors. If the pressure were not controlled, it would drop as this happened and negate the effect of opening the ports because the injectors are designed to operate only at the higher pressure. Therefore as the engine's need for fuel increases and the metering ports are opened, the pressure-regulating system adjusts so that pressure at the injectors remains the same.

Metering the Fuel

As air rushes into the engine, the delicately balanced sensor plate lifts up, moving up the control plunger which opens up the metering ports for the four injector lines. This part of the system reminds me of the SU carburetor. The throttle butterfly opens, but more fuel is delivered only when a plunger (SU) or a sensor plate (C.I.S.) is sucked up enlarging the orifice of the metering jet. The slight lag while this happens eliminates the lean fuel-mixture period that occurs in conventional carburetors, and that requires an accelerator-pump jet to compensate.

The sensor plate, and therefore the car, will not work properly if there are air leaks in the connecting air duct from the air-flow sensor housing to the throttle-butterfly housing. The sensor plate (and control plunger) should move freely and with uniform resistance throughout its range of travel. This can be checked by disconnecting the air ducting to expose the sensor plate. The edge of the sensor plate close to the fuel distributor should be aligned with the narrowest part of the venturi. Pick up the plate with a magnetic tool and feel its up-and-down movement. Any problems with the control plunger or fuel distributor are solved by replacing both as a unit. The gasket at the bottom of the fuel distributor should be renewed as well.

Mixture and Idle Speed

The idle mixture is set by first removing the rubber plug with a wire handle near the fuel distributor. This exposes a screw which is reached by a long skinny screwdriver. A turning of the screw changes the mixture by adjustment of the rest position of the control plunger. Counterclockwise movement leans the mixture. The idle speed is controlled by a bypass screw similar to that in the older fuel-injection system. Extra air for faster idle on a cold start is controlled by an auxiliary air regulator also similar to that in the older fuel injection system.

Fuel-System Checks

The main thing to check in a C.I.S. fuel delivery system is the fuel pressure. The dealer uses a special gauge and as of this writing nothing comparable is available to the general public. If you get one of the special banjo fittings from the dealer, you can attach and clamp a conventional fuel-pressure gauge to it (making sure it measures at least up to 60 p.s.i.) and insert it in place of one of the injector lines. This will allow you to measure the regulated pressure (which is the most important, anyway) as well as the residual pressure when the ignition is switched off. Full pressure from the fuel pump cannot be measured by this method, but it will fail by not working at all in most cases.

When the engine is cold, the pressure should range from 14 to 35 p.s.i. When the engine is warm, the pressure increases to 48 to 54 p.s.i. If this warmup effect is not observed, check the pressure regulator. Turn off the ignition and wait 10 minutes. If the pressure has not dropped below 25 p.s.i., the system probably has no leaks. If it does, check for leaks in the fuel lines, the fuel distributor, the injectors, the cold-start valve, and the fuel pump, especially the fuel-pump check valve. The fuel-pump check valve is a separate replaceable item screwed into the outlet of the fuel pump.

You can check the injectors by pulling them out of the manifold tubes where rubber seals hold them in place. Don't disconnect the fuel hose. Remove the fuel-pump relay and bridge terminals L13 and L14 with a jumper cable. Turn on the ignition without starting the car. Then pick up the sensor plate with a magnetic tool and observe the spray pattern. It should be even and cone-shaped. Turn off the ignition and see if the injector drips fuel. If the injector drips or has an irregular spray pattern, check the rest of the injectors and replace the faulty ones. Soak the rubber seal in gasoline upon replacing the injector.

Electrical Components

The C.I.S. has three electrical components: the connection to the fuel pump, the control-pressure-regulator, and the cold-start valve. When the ignition is on, you should be able to measure battery voltage with a test lamp at one of the prongs of the connector going to the fuel pump and the pressure regulator. If not, trace the electrical system backward to find the source of the problem. The fuel pump is activated by a fuel-pump relay. The two male prongs on the pressure regulator should give a reading of around 20 ohms when measured by an ohmmeter. If it doesn't, replace the unit.

The cold-start valve doesn't operate unless the thermo-time switch screwed into the head completes the circuit to ground. This only happens when the engine is cold and for approximately 8 seconds after the starter is operated. You can check it by carefully disconnecting the plug from the cold-start valve and putting a test lamp across the the two connectors in the plug. Pull the high-tension cable out of the distributor so the car won't start. Operate the starter continuously for around 15 seconds. The test lamp should go on, then go out after 8 seconds or so. The prong in the plug for the cold-start valve, which is "hot," is only hot when the starter is operated. If it is hot at all times when the ignition is on, the wires on the starter solenoid are confused. The lead for this valve should come from terminal 50 rather than from terminal 30. The cold-start valve itself should measure below 3 ohms between its terminals. More information on this system is available in the excellent *Rabbit Service Manual*, published by Bentley.

Chapter 6:

Smog Devices

To live outside the law,
you must be honest.

—Bob Dylan

Most of the basic machinery of an automobile is the product of decades of evolution; of thousands, if not millions, of trials and errors. The evolution of the engine itself has experienced the molding forces of the need to achieve better performance and increased economy. The desire to control the nature of the waste products of the automobile introduces an entirely new direction of change. The variety of devices used to control engine waste remind me of the crazy inventive days before General Motors, Ford, and Chrysler standardized the automobile and made it Big Business; the days when a couple of brothers with a few thousand could start an automobile business.

Every modern car type has its own solution to waste control. The solution is constantly changing, and some of the devices do the opposite of what they were intended to do. Often a good tune-up is all that is necessary for the car to pass the smog-emissions test for that year and type of car, but devices still have to be added to satisfy legal requirements. Nevertheless, the basic intent of devices and laws to control exhaust is laudable, indeed necessary if the automobile is to continue in massive use. The future will probably bring simpler solutions to exhaust control.

The following discussion will cover the general approaches to exhaust control, but the specific details for each car have to be left to a manual for each particular car. *Chilton's* is also a good source for the car-by-car, year-by-year details. Sometimes the only place to find an explanation for some of the more short-lived smog devices is in the handbook which came with the car.

Positive Crankcase Ventilation (P.V.C.)

The first exhaust products to be controlled were the crankcase funes. The crankcase has always had to be vented to keep pressure from building up. The pressure came from gases blowing by the rings on the compression stroke. This happens even on a new engine. On an engine with broken rings, oil will actually be blown out of the breather tube. The breather tube on older cars simply vented down below the engine into the airstream which passes under the car when it is moving.

The simplest control of this exhaust product was to connect the crankcase vent to the air-cleaner. The fumes would then pass through the air cleaner and become part of the mixture burned in the cylinders; air cleaners got dirtier sooner.

Another method of exhaust control was to modify it from a passive system of fume collection to an active closed system. One approach was to connect a hose to the valve cover through a PCV valve to the intake manifold. The vacuum in the intake manifold would suck the fumes from the crankcase and the PCV valve would prevent flow in the opposite direction. This would keep a back fire in the carburetor from causing an explosion in the crankcase. Clean filtered air would come in through the air cleaner and through a hose connected to another portion of the valve cover. In earlier versions fresh air simply entered the oil-filter cap. The crankcase has a common air space with the valve area of a push-rod engine. This is not the case with some overhead-cam engines.

If there is a vacuum leak in this smog-control system or if the PCV is the wrong one for the car, the car may idle badly or not at all. If the system gets clogged, pressure can build up in the crankcase and oil leaks occur throughout the engine. To check the PCV valve for blockage, give it a shake. If you hear the jiggle pin jiggling, the valve is probably OK.

Smog Pump

The next device to come along of the common exhaust-emissions controls is the so-called smog pump. Look Ma, an afterburner! What we have here is an air pump run by the engine, which injects air into the exhaust manifold to assist in the continued burning of unburned hydrocarbons. Unburned hydrocarbons are more noxious to humans than burned hydrocarbons.

The air-injection pump is found near the alternator in most cars and is operated by a fan belt. It can therefore suffer all the problems of fan-belt operated machinery (see *Electrical III*). An intake hose connects with an air filter (either the carburetor filter or a separate one), to allow fresh air to be sucked in. The pump itself is designed to produce a pressure of 1 p.s.i. at the injectors. It has a relief valve to protect the pump against pressure buildup. This valve is in the body of the pump and is at times the source of a horrendous rattling at idle speeds. This does not necessarily indicate a malfunction, but it is the one part you can replace on this unit. Leading from the smog pump is another rubber hose, of large diameter, which goes to a check valve. That valve is screwed into the manifold, which distributes air to the various exhaust ports in the head. On most cars an antibackfire valve is fitted between the smog pump and the manifold. Sometimes the air manifold is cast as part of the head, in which case the check valve screws into the head. The check valve prevents exhaust from going back up the system to the antibackfire valve or the pump. The valve gets corroded from the exhaust and therefore definitely doesn't last forever.

To check the check valve, disconnect the air hose running to it and, with the engine running, see if there is an exhaust leak through it. If exhaust gets back to the pump, the bearings will freeze and the pump, of course, stops working. Air pumps are usually fixed by replacement. Exhaust manifolds, by the way, run significantly hotter with this afterburner setup.

Antibackfire or Gulp Valve

When the manufacturers added the smog pump to the automobile they found that an antibackfire valve also had to be devised for the special case when the engine decelerates from high speed. Extra unburned gas gets dragged through the engine into the exhaust manifold, which if mixed with oxygen will cause backfiring. Backfiring is unacceptable even if it burns the extra hydrocarbons. Therefore air is shunted away from the injection system to the air cleaner in some cars, or the intake manifold in others, or simply to the atmosphere. The action of the antibackfire valve is controlled by vacuum in the intake manifold. The vacuum connection between the two is a small-diameter rubber hose. If backfiring is a problem, then this antibackfire valve may be at fault.

To check this valve, disconnect the large-diameter air hoses which go to the check valve. (The input is the hose from the air pump.) With the engine idling, air should be coming out of the fitting, which is normally connected to the check valve. Pull the accelerator linkage to get a fast idle (2,000 r.p.m.), then let the linkage snap closed. The air should momentarily be diverted from its normal output connection to the other output air fitting or directly to the atmosphere. All hoses should be checked for leaks when they are hooked up.

Exhaust-Gas Recirculation

While the smog pump injects air into the exhaust manifold, the EGR system bleeds small amounts of exhaust gases into the intake manifold. In this way the nitrous-oxide emissions are reduced by lowering the combustion temperature. The device is sometimes an addition required for older cars. The device is only supposed to be activated at middle and upper engine speeds. Therefore it is controlled by vacuum from the venturi section of the carburetor, which is near zero at idle and goes up with increased air rush.

If you wish to check this unit, temporarily connect the vacuum connection to the intake-manifold vacuum while the engine is idling. Vacuum in the intake manifold is highest at idle. This will cause exhaust fumes to vent into the carburetor. The engine will idle badly or even stall if the EGR system is working. The greatest problem with this unit is that corrosive exhaust fumes plug it up and make it inoperative often within six months.

Actually high-performance engines with high exhaust and intake-valve overlap due to the cam design, have a fair amount of exhaust contamination in the fuel/air mixture. This is desirable at high engine speeds because the carbon monoxide in the exhaust retards burning and therefore preignition. It gives the high-performance car a terrible idle, however.

Thermostatically Controlled Air Cleaner

The function of the choke in the carburetor is to increase the fuel portion of the air/fuel mixture. This increase compensates for the fact that with a cold engine and cold air rushing into the carburetor, the gas does not atomize well and tends to condense on the intake manifold and in the cylinder. The thermostatically controlled air cleaner shortens the time of the cold-air condition, and the choke can turn off sooner.

Since the exhaust manifold gets hot rapidly, heat is collected by a stove device around the manifold and is vented by a special insulated hot-air hose to the air cleaner. By means of a bimetal spring the air cleaner controls the direction of air flow. When the temperature is at 100 degrees and above, the damper door should be open forcing all the air to come from the outside. Air from the manifold stove should be too hot at this time causing the fuel mixture to become too lean.

There is another vacuum connection to this device: a small-diameter hose running from the intake manifold to the air cleaner where the air bleed valve is located. From this valve a hose runs to the damper door. The function of this vacuum connection is to open the damper door temporarily and let outside air in when the accelerator is snapped closed with the engine running at higher r.p.m. When the engine is warm and the damper door is open to outside air, the air-bleed valve, also being controlled by a bimetal spring, closes off the vacuum connection to the damper door.

If there is an exhaust-manifold leak in the area of the heater stove, exhaust fumes will be carried up to the air cleaner. This will cause the air cleaner to get dirty and clogged in a few thousand miles. The car will then idle poorly if at all. Removing the air cleaner should allow the idle to improve, if this is the cause of the problem.

Lean Carburetor

Engines run somewhat inefficiently at idle because of the high vacuum in the intake manifold. Since there is a certain amount of valve-opening overlap (the intake valve and exhaust valve are open at the same time for a short period), exhaust gases remain in the cylinder contaminating the incoming fuel charge. This overlap is a result of valve timing built into the design of the cam, and it produces better performance under most conditions. High-performance cams have even more valve overlap and an even worse idle than the family-car touring cam. This idle insufficiency is reflected by a need for a richer 12:1 fuel ratio rather than the normal 14:1 ratio that is sufficient at cruising speed. This richer setting, however, leads to more unburned hydrocarbons.

Setting the idle-mixture jet lean leads to more complete burning of the fuel. The car will no longer idle reliably hence the idle speed has to be increased (the throttle is opened a bit). This increase will turn out to be much too fast, so the next step is to retard the spark, which slows the idle again. With the timing retarded as much as 5 degrees after top dead center, the engine is dreadfully out of time but at least it idles slowly. This seems to be a case of two wrongs making a right. The timing is returned to a more normal setting when the car is moving. This return to normal is controlled by various devices. (Consult shop manuals or *Chilton's* for specific details.) Special vacuum units are used on the distributor. Some have a double diaphragm, some a single. They all require a vacuum connection to the intake manifold to activate the retard section of the vacuum advance.

When the carburetor is set lean, the engine runs hotter. To correct for overheating, a temperature-sensing device shunts vacuum to advance the spark and speed up the engine, thereby cooling it down. Now we are nearly back where we started, I'm afraid, but that's the way it goes. Pay attention now! Things are going to get more intense.

Higher-Temperature Thermostat

Inspired by some theory of thermodynamics and good logic, some car manufacturers put higher temperature thermostats in the cooling system, which allows the engine to run hotter. This, in practice burned excess hydrocarbons and thus reduced harmful exhaust emissions. Unfortunately, a new type of harmful exhaust emission was discovered that was produced by the hotter engine. It turns out that the type of chemical reaction, therefore the by-products (exhaust), depends on temperature, pressure (varies with engine speed), gas used, and fuel/air ratio at least. In other words, the probability of discovering a nonpolluting internal combustion engine by this approach is vanishingly small. All the various types of emissions have yet to be discovered. We do know that the burning of one gallon of gasoline produces one new gallon of water. I wonder if this will ever bring us to grief.

At any rate, we now have hotter running engines. These required changes in the cooling systems. A new fluid was invented to replace water. It has a lower freezing point and a higher boiling point than water. Both properties are desirable especially for smog-controlled engines. Unfortunately, now we have this new liquid, which, although it may replace the water in the engine, eventually will become part of the water we drink.

In another change, the cooling system has been designed to operate at higher pressures and therefore has to be tighter. This helps save fluid, and higher pressures raise the boiling point of whatever fluid happens to be in the system.

Other Changes

The timing has been fooled around with endlessly. Let the explanations go in one ear and out the other; just smile and set it to the values recommended. If the car still doesn't work, set it to the place where it will work.

Another craziness has been the throttle damper, whose function was to hold the throttle open during deceleration to burn the excess gas sucked through the engine under this condition. This changes deceleration into something more resembling acceleration unless the clutch is put in or the brakes applied. The VW people call their device "throttle valve regulator". In wide experience with the VW, I have only seen one that was hooked up and worked. I can adjust the device to work properly for a few days. Electronic fuel injection is replacing conventional carburetion because the latter supposedly does not control excess gas during deceleration.

I think that something stuck on the exhaust pipe, perhaps water cooled to the inside heater, has the best chance of improving the exhaust-emission problem. The catalytic converter on the exhaust line is just such a system, but it has its drawbacks. If it gets overworked by an out-of-tune engine it will overheat. If an additive in the gas doesn't suit it, the converter will clog up. This will give poor engine performance because of reduced engine breathing. The reduced performance can be identified with a vacuum gauge. Hook the gauge to the intake manifold and start the engine. If the vacuum reading slowly drops to zero when the engine is idling, the exhaust line is clogged.

General Trouble Shooting

The biggest problem encountered with smog equipment in the course of normal maintenance is vacuum leaks. If the little vacuum hoses are loose or not put back, the car will not idle at a reasonable speed or will overheat.

If it is doubtful whether the engine and its normal accessories or the smog equipment is causing a particular symptom, it is usually easy enough to disconnect the smog devices from the engine. You should be able to get the engine to run at least as well, often better, with the smog-equipment disconnected. If the symptoms persist, look to the engine for problems.

The Vacuum Gauge

The vacuum gauge is a useful diagnostic instrument for quickly checking the integrity of the entire vacuum system. The normal car should show from 17-22 inches of vacuum at idle. A leak might give a reading of 4-9 inches of vacuum. A slight vacuum leak might cause a drifting needle on the gauge. A vacuum gauge also indentifies a sticky valve or a blown head gasket. A list of diagnostic indications comes with the gauge.

To narrow down the source of a vacuum leak, attach the vacuum gauge to the intake manifold and crank the engine without starting it (ground the high-tension cable going to the center of the distributor cap). With the air cleaner removed, place your hand tightly over the mouth of the carburetor. The normal reading should be around 17 inches of vacuum. If it is low, remove the carburetor, block the intake manifold, and repeat the test. If the reading is still low, the leak is in the intake manifold or at the manifold gasket. If the reading improves, the leak is at the carburetor, possibly the throttle shaft, or one of the vacuum connections. To squirt oil around areas of potential leakage might pin down the problem. A vacuum leak will cause the oil to disappear at the leak.

Most vacuum hoses will be alright without the use of clamps, if the hose is the right size and not dried and cracked.

Chapter 7:

Electrical Systems I: General Overview

And note also, that in this Discourse I do not undertake to say all that is known, or may be said of it, but I undertake to acquaint the reader with many things that are not known to every angler.

—Izaak Walton (1593-1683)

This chapter will discuss basic electrical concepts and trouble shooting concerning the car's electrical system. Three more chapters will focus on more specific areas: the battery and starting system, the ignition system, and the charging system. First let me clarify some basic electrical terminology.

Electricity is often compared to water flowing through a pipe. The size of the pipe determines the maximum amount of water which can flow with a fixed pressure pushing it. Similarly, the diameter of the wire determines the maximum volume of electricity, the *current*, which can flow through it, if the pressure remains constant. The pressure pushing it through the wire is the *voltage*. A 12-volt system has more pressure than a 6-volt system. Since work done by elec-tricty depends on both the volume of electricity (current) and the pressure (voltage) behind it, a 6-volt system requires twice as much current to do the same work as a 12-volt system. The wiring in a 6-volt system has to be heavier than in a 12-volt system to carry the extra current.

Opposing the flow of current is *resistance*. Sometimes the resistance is inherent, such as the resistance that is in a wire. Copper has less inherent resistance than aluminum, hence it is better for electrical conduction. Silver or gold have even less inherent resistance than copper, but are too expensive.

Resistance may also be accidental, such as oxides on switch contracts which impede electrical flow, or corrosion on battery posts which does the same thing. The resistance on the battery post can be so great that it impedes any current flow. This resistance can be reduced by cleaning or replacing the parts.

Finally, resistance may be planned, such as installing the ballast resistor in the ignition system of many 12-volt cars. Its function is to reduce the voltage on the current going to the coil to 8 or 9 volts when the car is running normally. The full voltage pressure is applied to the coil during starting to get an extra hot spark. The coil, which may actually be an 8-to-9-volt coil, would overheat and have a rather short lifetime if run continuously at 12 volts. Also, the points would burn up sooner.

Damage to an electrical system because of poor connections or shorts can usually be attributed to heat. One of the most annoying results of overheating is a wire that shorts to ground drawing more current than it was designed to carry. It gets hot, melts its insulation off and, if it happens to be wrapped in a bundle with other wires, melts through their insulation. The next thing you know, you switch on the turn signals and the windshield wipers start to work because wires have shorted to one another.

It is essential for any detailed analysis of the electrical system that the mechanic has, and understands the use of, a voltmeter (measures volts), an ohmmeter (measures resistance), and an ammeter (measures current).

Voltmeter

The simplest form of voltmeter is a test light. It is nothing more than a light bulb and a test lead. It tells you whether a wire or connection is "hot" or "cold" that is, whether there is electrical current moving in the wire or not. If your taillight is out and you don't know whether the bulb has blown or whether no current is getting to the bulb, you can remove bulb and see with the test lamp if the contacts in the bottom of the bulb socket are hot. If they are, the bulb has blown.

The brightness with which the test lamp glows gives a rough idea of the voltage in the wire, but a more accurate measuring device is the voltmeter. For car work a voltmeter, which has a scale of 0 to 30 volts and a scale of 0 to 3 volts, is most useful. The voltage going to most parts of the car, that is, to the various lights and electrical motors, is the same as battery voltage (6 or 12 volts). Usually only the voltage going to the ignition system is purposely reduced by inserting a resistor "in series." The word "series" introduces some basic concepts that have to be defined.

There are two basic wiring principles. In *series* wiring a wire connects the battery to a string of resistances and then returns to the battery.

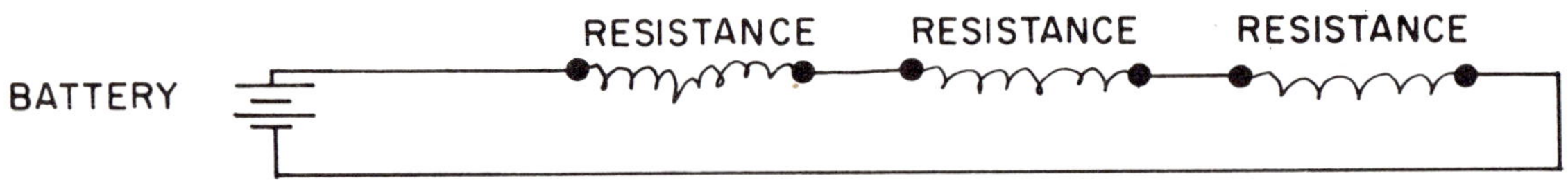

The resistances may be resistors or they may be light bulbs or motors that have reisistance. If there is a break in the line anywhere or if one bulb burns out, then nothing will work. Some cheap Christmas-tree lights are wired in this way. The fuses in a car are wired in series with the parts they are protecting. If there is a short in the system causing a large draw of current, the fuse heats up, melts, and all elements in the curcuit after the fuse will go dead for lack of current. To test a fuse, turn on the ignition, attach a test lamp to a good ground, and touch the probe to each side of the fuse. If both sides are hot, the fuse is OK. If you were to hook a voltmeter in series with the circuit above,

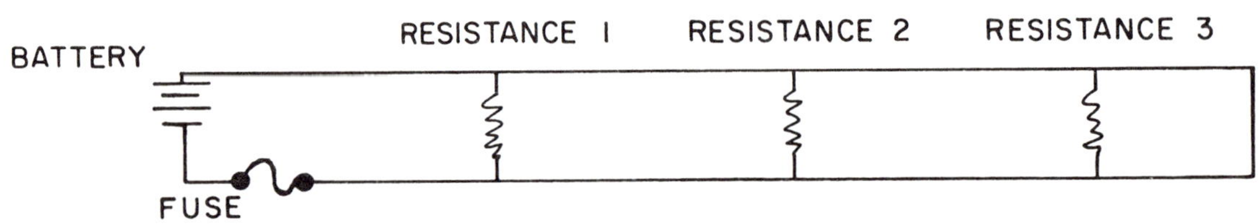

it would be measuring the total electrical pressure in the line and it would be one of the resistances in the circuit.

The other basic wiring principle is *parallel* wiring. In this case, all resistances in the circuit are independently wired to the battery.

Most of the car's wiring is parallel. The drawing above could represent the wiring of the left taillight (resistance 1), the license-plate light (resistance 2), and the right taillight (resistance 3). If one of the bulbs goes out, the rest of the bulbs will still work. If the fuse is blown or if there is a break in the wire running from the fuse to the back of the car, then all lights will be out. The diagram shows a wire running back to the ground side of the battery, but in a car the body itself makes up this part of the return circuit. The final step to the battery is completed by a ground strap either attached to the engine or the body. The engine should have a ground strap to the body so that it becomes part of the ground side of the circuit.

Hooking the voltmeter in parallel with some part of the circuit is a useful way of measuring resistance in the circuit if current is flowing through it. This technique is called measuring the voltage drop. The greater the resistance in the circuit, the greater will be the voltage pressure bypassing through the meter. For example, hooking the positive (red) lead of the voltmeter to the positive post of the battery and the negative (black) lead to the battery or positive post of the coil we get a voltage reading if the ignition switch is on. If there is no ballast resistor in the ignition circuit, the voltage reading will normally be under 1 volt. If there is a ballast resistor in the circuit, the voltage reading will normally be from 4.5 to 7.5 volts depending on the value of the resistor and the temperature. (Resistance is affected by temperature.) If the readings are higher than normal, then additional resistance is in the circuit in the form of poor connections or bad switches.

The voltage drop between the battery post and the starter terminal should be 0.6 volts or less (The starter system must be operated to make these tests). If it is more, shift the positive lead from the battery post to the battery clamp. If the voltage reading drops to normal, excessive resistance exists between the battery post and the clamp. The voltage drop between the negative battery post and the starter-motor case should be 0.2 volts or less. If not, the grounding is insufficient.

It might be thought that measuring the voltage directly across the battery posts would give an accurate measure of the battery condition; that on a 12-volt system a reading of 10 volts or less, ignition off, would indicate a bad battery. But this is a rare case; usually a battery will still be reading around 12 volts even near the end of its life.

A more accurate way to think of battery efficiency is in terms of resistance. A good battery has low internal resistance to current flow. Therefore, a measuring of the internal resistance would give a good estimate of a battery's health. This is most easily done by measuring the voltage drop when the battery is under a load. The battery has to be under a load to get a large current flow. Resistance cannot be measured without current flow.

Hook the voltmeter leads to the battery, red to positive, black to negative. Hook a jumper wire from the distributor lead to the coil to ground to keep the car from starting. Crank the engine about 15 seconds. The voltage reading for a healthy battery will be 9.6 volts or higher (12-volt system) and 4.8 volts or higher (6-volt system). The internal resistance in the battery is implied by the voltage reading. Unfortunately, if the reading is low the cause could be also excessive resistance in the starting system. Peerless makes a battery-testing voltmeter with a built-in resistance that duplicates the resistance load of a healthy starting system.

Ohmmeter

A voltmeter is not the only tool for measuring resistance. The simplest resistance tester is a continuity tester. This is a bulb with a test lead and probe similar to the test lamp except that it includes a battery. This means that it will introduce its own small current into a circuit to measure its electrical continuity, whether there are any breaks in the line or not. A break in a line would be equivalent to infinite resistance, and the light will not go on. The ignition switch should be off when this tester is used because the car current will blow the bulb in the tester. Have a few spare bulbs around for times when you forget. Also check this tester each time before use by momentarily touching the two leads together to see if the bulb lights. Like flashlights, this instrument always has corrosion problems.

A continuity tester determines whether there is a break in an electrical circuit or whether a switch is on or off, but if you wish to measure exactly how much resistance is in a system, you need an ohmmeter.

An ohmmeter uses a battery. Since the strength of the battery influences the meter reading, it has to be re-zeroed with every use. It should be turned off, or at least the test leads kept separated, so that the battery isn't drained when it is not in use. For car work it is useful to have two resistance ranges on the meter, one range from 0 to 1,000 ohms and the other from 0 to 10,000 ohms, approximately.

The ohmmeter may be used to measure any electrical windings. Measuring the windings in the coil (see *Electrical Systems III,*) along with a spark-gap test is sufficient to determine the health of the coil. There are a dozen induction sensor coils, electrical switches, motor windings, and resistance temperature-measuring sensors in an electronic fuel injection system which can be measured for good health quickly with an ohmmeter (see *Fuel Delivery III,*).

The limitation of this instrument is that one has to know what resistance values are normal for various electrical windings or wires. This is often hard to know. Some values will be found for various electrical devices in this book under the headings *Tuneup, Carburetion, Electrical II, III,* and *IV*.

Ammeter

Older cars and those which pride themselves on being fully equipped have an ammeter on the dashboard. Your average contemporary car has a red light instead of an ammeter. This is supposed to go on when the ignition is turned on but off again when the car is running.

Small induction ammeters are available at any good parts store. These can be placed along a wire to measure the current flowing in the wire. It's magic. If the car is electrically healthy, the ammeter, if placed alongside the battery cable, reads as follows: The needle will stay centered until the ignition switch is turned on. It will then move ever so slightly to the left indicating a slight current discharge through the grounded points of the ignition system. If the lights or other accessories are turned on, the discharge will be greater. The normal discharge for the points being closed, with no other accessories on, should be approximately 1½ amps (12-volt system). If it is more, suspect a defective ignition system or ignition switch. All accessories may be checked by individually switching them on. The permissible current is found by dividing the wattage of the accessory by the voltage of the system. For example, the headlights are each 40 watt bulbs; divide 80 watts by 12 volts and you get the answer 6.67 amps. Therefore, when the headlights are switched on, the ammeter should show approximately 6.67 amp discharge. If it is more, resistance is excessive, probably at the connection points. If the needle goes all the way left, there is a dead short to ground somewhere. If no discharge is seen, there is an open circuit. When the ignition key is twisted to the start position, the greatest discharge should be seen as the starter solenoid and starter motor are added into the circuit.

Once the engine catches and the car runs at fast idle, the ammeter needle should swing sharply to the right. For a moment it should show the maximum output capacity of the generator and alternator. This ranges from 30 to 50 amps. If it goes to its capacity, the generating system is healthy. The peak is soon lost and the needle will gradually vibrate down to a value just to the right of center as the battery becomes recharged. If the needle acts very erratically, the points in the regulator are probably sticky. For a more certain check of the charging system, turn on the headlights for five minutes before you start the car, to run down the battery a little. If the ammeter doesn't show charging at the rate at which the generator or alternator is capable, those units are perhaps in need of service. None of this information is available to you for the idiot light. Further discussion of this matter will be found in *Electrical Systems II*.

Electrical Connections

We are coming to the yin and yang of electrical connections. The male side of any electrical connection is a prong, a spade, or something sticking out. This prong is electrically dead until it receives current from the hot female receptacle. Making the female connector the hot lead is a universal practice. Most connectors used by manufacturers are the crimp-on spade type nowadays, except for heavy wires. Soldering is still best, but if you test your crimp connections by trying to pull them apart and rejecting those that do, the rest will be acceptable. Many mechanics twist wires together and call it a connection. If such a connection is found, it should be improved. The most critical electrical connections in the car are the battery-post clamps (see *Electrical Systems II*).

Electrical Switches

The starter switch is basically a three-position switch—off, on, and temporarily on. The temporarily-on position is spring loaded, so it stays in this position only if it is held there. This switch, like all others, can build up deposits on the contacts. The deposits are usually due to burning, which can also be a problem with the points (a switch in the distributor), and the relay contacts of the regulator (a switch in the regulator). The switch contacts in the distributor or the regulator can be filed and then burnished, a process of polishing by rubbing, but the ignition switch is repaired by replacement.

Some other on/off switches oxidize when they are not used. This problem, which is also common on hi-fi equipment, is solved by turning the switch on and off 20 to 30 times to polish the contacts.

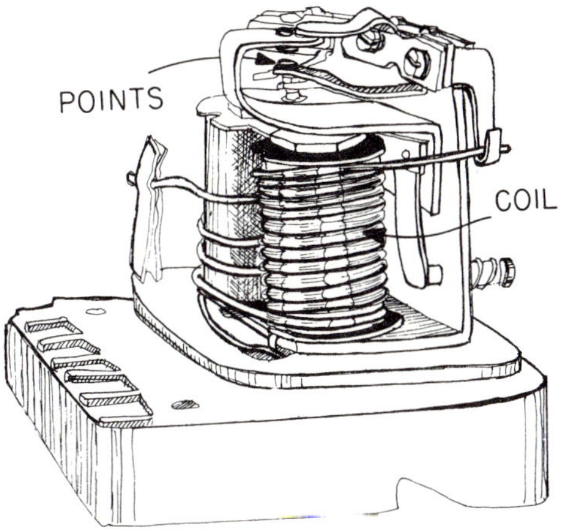

The relay is an interesting automatic switch. Anything you wish to turn on and off is wired to the contacts of the relay. Under the contacts, which could either be normally open or normally closed, is an electromagnet. When it is activated, the contacts either open or close. The voltage-regulating relay in the regulator controls the field circuit of the generator. When the voltage level of the generator output reaches a certain point, it produces enough of an electromagnetic field in the relay coil that it overcomes the spring tension tending to keep the switch closed or on. Thus it turns off the feedback to the generator field coils causing the generator output to fall. This reverses the process, the relay points close again, and the field coils are re-energized.

If two or more horns are used in a car, they are not directly activated by the horn button. The horn button usually closes a relay which completes the horn circuit. Relays are used to avoid the running of many heavy wires with heavy duty switches into the driver's area. The headlight switch is an exception to this, but not because it isn't a potential source of trouble—it is.

Another long-distance switch is the starter solenoid. It is a switch that has to carry up to 200 amps at times. The contacts of this switch are very large. Sometimes this switch can be cleaned up by reversing the contacts.

Some dashboard switches can be tricky to remove. The commonest mounting requires the pulling off or unscrewing of the knob of the switch. Sometimes a small set screw or a recessed Allen set screw has to be loosened before the knob is removable. Some knobs on early English cars required the pushing through of a hole in the knob to depress a little catch before the knob was pulled off. When the knob is off, a large nut or a chrome slotted ring must be screwed off. The chrome ring may be loosened and nudged around with a screwdriver, while the end of the screwdriver is hit with the heel of your hand. When the ring is removed the switch can be pulled out from behind the dashboard.

Some knobs are permanently attached to the shaft of the switch. If this is the case, the knob and shaft come out together. A catch in the body of the switch holds it in place. The catch is released when a thin metal rod is pushed through a hole in the body of the switch under the dashboard and the shaft is pulled out or a button on the switch body is pushed.

Tracking Shorts

When looking for shorts, first inspect the fuses. This can often narrow down the area of the problem as well as explain a malfunction. Don't insert fuses that are much higher than the recommended value. This could lead to overheating of the wires and the need to replace a wiring harness. A wiring harness is a bundle of wires taped together, all color coded, of the correct size and with the appropriate fittings on the end.

The whole wiring system in many American cars is protected by a "fusable link." This is a soft metal wire which runs from the battery to supply current for the car's accessories. If the car is completely "electrically dead," check this "fusable link" and replace it if necessary.

If the vehicle has been converted from anything to anything else, especially if wiring splices were used, check the rewired areas at the first sign of electrical problems. Remove electrician's tape from a wire in a car, and nine times out of ten you will find wires twisted together. Often even wires spliced together with crimp connections, if exposed to the weather, will develop an oxide on the base ends which reduces the current flow.

I remember a converted school bus I worked on once; the dash warning light would turn on and the vehicle would run if jump started, but the starter solenoid chattered when the key was held in the start position. This solenoid chatter often results from a low battery voltage or a corroded battery cable. After I eliminated these potential problems areas, a test light showed that current was flowing through all the switches and a firewall connection. Then I noticed a splice in the hot wire going to the ignition switch. I pulled on it and it came apart; the ends were corroded. Had I measured the voltage drop between the battery post and the solenoid, it would have shown excessive resistance in the circuit. Enough current could flow to light the generator and oil-pressure lights, but not enough to operate the solenoid.

Open circuits are the least troublesome. Usually something doesn't work because it is disconnected or a fuse is blown. High resistance in a circuit is more insidious. Motors that don't receive their full requirement of current will often still run, but run hotter and have a shorter lifetime. Circuits shorted to ground cause overheating and burning. The burning starts at connection points, always the weak points of a circuit, and may proceed rapidly along the whole length of the wire affected by the short.

One of the worst instances of shorting to ground is when the ignition wire in a VW bus gets hot. This wire is part of a large wiring harness, which runs from the front of the bus to the rear through the frame. It includes 14 wires, each color coded and in three or four different diameters. It would be impossible to find anything nearly the same for a lesser expense if one started from scratch. Of course, you could get a big roll of red wire and hang replacement wires underneath or run them through the inside: but the worst wiring repairs, even worse than replacing a wiring harness, are encountered in such homemade shortcut rewiring jobs. Whether you are a mechanic or a consumer, I recommend you stay away from such vehicles.

There is a basic trick to replacing difficult wiring harnesses. You temporarily tie the ends of the old wiring harness together with the ends of the new one; the rear ends of the old to the front ends of the new. Wrap this area tightly with electrician's tape and then carefully pull the old harness out from the front. The new harness gets pulled through at the same time. A horn wire that runs down the steering column and out the end of the steering box on many cars is often difficult to replace except by this method. There are cases where you can use a long piece of 26-gauge brass wire to thread through the wire channel, tying the end to the new wires and pulling them through from the other side.

Malfunctioning Lights

The most probable cause of lights being out is a blown light bulb. Any bulb can be checked by attaching two alligator test leads directly from the two battery posts, one to the bulb base and the other to the center contact.

The next most probable cause of lighting problems are broken taillight lenses, or body damage which keeps the light socket from being protected against weather. The screws which hold the light fitting in place also complete the electrical connection to ground. Sometimes unscrewing and screwing these fasteners improves the ground connection. I also spray the area with WD-40 to clean it up and to remove moisture.

If the turn-signal light doesn't flash for left or right turns, it is usually the result of a bad flasher. Both the front and rear lights have to be working for the flasher to function, because only then is enough current drawn through the flasher unit to make it work. Therefore, if the turn signal works in one direction but not the other, one of the turn-signal bulbs is probably out. If the turn signals turn on without flashing, perhaps a different flasher is required.

Electrical gauges by and large operate on a principle of variable resistance. A change in temperature, pressure, or gas level is translated into a change in resistance. The needle on the gauge is often situated between two coils.

Gauges

When the resistance in the sending unit screwed into the block, or in the gas tank, changes, the current flowing in the coils changes proportionally and the needle moves. These simple gauges are quite long lived. Some Chrysler and Ford cars have a voltage regulator, which keeps the voltage down to a constant level. If all gauges fail at once, this regulator is the prime suspect.

If a single gauge fails the sending unit is probably bad or the wire connecting the sender with the gauge. Often you can see physical damage on the sender. If it is an oil-pressure sending unit, look for oil leaking from the sender. The gas-tank sending unit can be removed from the tank and examined for free movement of the float. On many newer cars the fuel-pump and gas-tank sending units are a single unit.

The gauges most susceptible to problems are generally those which are nonelectrical, the speedometer and the Bourbon-type gauges. The Bourbon-type gauges are similar in principle to the gauge used for measuring the compression vacuum or gas-line pressure. Bourbon-type temperature gauges are used on many English cars. They are sensitive and accurate, registering temperature at all times (ignition off). Their drawback is that they are delicate. The sending unit and gauge are an integral unit connected by a capillary tube filled with ether. If this tube leaks, the whole unit has to be rebuilt or replaced. A speedometer shop can do this.

Bourbon-type oil-pressure gauges are also nice to have for accurate, fast reading, but introduce more high-pressure oil-system plumbing which, of course, can leak.

Stop-Light Switches

The stop-light switch, which screws into the master cylinder, is prone to leak fluid through the switch after a while. The switch probably won't be working when this happens. When replacing this switch have the new one in hand to screw in immediately as the old one is removed so you won't lose brake fluid. Before tightening the switch all the way, pump the brake pedal a few times until the "spitting" stops, then quickly do the final tightening. This helps to get air out of the brake line. Some cars, like the VW, have brake-system warning lights. These have a "failsafe" switch, which can malfunction and cause brake failure. I wonder if it was worth introducing these lights.

If there are no wires running to a pressure switch on the master cylinder, look for an on/off switch placed near the brake pedal inside the car. If a continuity light is attached to the two leads behind the switch and the button is pushed, the continuity light should turn on and off if it is working.

Horn and Turn-Signal Switch

Removal of the steering wheel requires a simple puller designed for this purpose. It is fairly inexpensive and available at most parts stores. Before we use the puller, the horn ring and hub have to be removed. Elaborate horn rings on some American cars are removed by the unscrewing of some Phillips head screws from behind the wheel. Simple horn hubs are of two types: a spring-loaded kind, which pops out with the aid of a screwdriver (VW and several other foreign makes), and a kind that you twist first (like a radiator cap).

When the hub is removed, a large nut is revealed, which has to be removed. Since the steering wheel is mounted on a tapered shaft, removing the nut is not sufficient to allow the wheel to be removed. The puller is required. There are often two screw holes in the hub of the steering wheel meant for the attachment of the puller. Before the steering wheel is pulled off, mark the end of the shaft in relation to the position of the wheel. There may be a mark already there. You will have to put the wheel back in its original position. The position of the spokes of the steering wheel is a useful guide to front-wheel alignment, but only if the steering wheel hasn't been moved from its normal position.

The horn wire may be part of the bundle of wires going to the turn signal or it may go through the center of the steering shaft and out the bottom of the steering box. In the latter type, a short in the wire where it comes from the steering box is fairly common. Use the old wire to thread a replacement wire through the steering shaft.

Unscrew the selector lever, and the turn-signal unit may now be unscrewed in preparation for removal. If the mechanical section of the switch comes in a unit with the wires, slip all wires at the switch and pull all but one of the wires out of the way. Do not disconnect the wires from the plug located somewhere under the dashboard. Use the remaining wire to thread the wire bundle of the new switch down the steering column. If this doesn't work, take a piece of 28-gauge brass wire and thread it up through the wire channel from the bottom, twist it around the new wire bundle, and pull the new wires through. At worst, you may have to thread the wires one at a time. Replace the old wires with the new at the connector, observing the color code. Now put everything together, reversing the steps used to disassemble.

Intermittent Electrical Problems

As a reflection of some of the mysterious malfunctions that can occur in the electrical system, especially if modifications are made, let me cite my present car as Exhibit A.

I modified the taillight on my Karmann Ghia from a two-bulb three-filament system to a single-bulb two-filament system. Don't ask why. When I now put on the brakes, the dashboard lights go on as well as the brake lights. Also, if the headlights are on and I turn off the ignition key and remove it, not only does the engine continue to run normally, but the horn starts to honk. Turning off the headlights turns off both the engine and the horn. This could be a nice way of reminding me that the headlights are on except that it doesn't always do it.

Intermittent problems are the hardest to diagnose unless the malfunction is occurring at the time of testing. If the problem isn't manifesting itself while you have your tools in hand, the best you can do is guess, which can be expensive. I frequently suggest the car be driven until the problem becomes worse or goes away.

Chapter 8:

Electrical Systems II: Battery and Starter Motor

The people that ceases to invent ceases to grow.
—Otis T. Mason (1838-1908)

If you get into a car, put in the ignition key, turn it to start and nothing happens (or perhaps just a few groans, then nothing), you have a problem with the battery-starter motor system. If the starter motor makes its normal starting sound but spins free before starting the engine with a whirring sound, the starter motor needs to be rebuilt. Probably the overrunning clutch is shot and needs to be replaced. This system also includes a starter solenoid, the ignition switch, a neutral safety switch with an automatic transmission, and various wires and cables.

First let's read the signs available from the driver's seat. If none of the idiot lights on the dash go on, then the battery is probably dead. Of course the failure of the idiot lights could be an open circuit in the wire bringing current from the battery to the ignition switch. If this is the case the headlights will still work on most cars. If the idiot lights do go on, then pull out the headlight switch. If the idiot lights dim significantly or go out, the battery has a low charge. Turn the ignition key back and forth a few times to see if you can hear a clicking sound. If you hear a single solid click each time the key goes to the start position, then the battery, the ignition switch, and the starter solenoid are working. If there is a rapidly repeating clicking sound, the battery is in low charge. The clicking comes from the starter solenoid, which is getting just enough current to close but not quite enough to stay closed. Therefore it rapidly oscillates between open and closed.

If the idiot lights go on and the headlights are strong, perhaps there is some special problem in starting this car. If the car has an automatic transmission, be sure the selector is in park or neutral. Jiggle the selector around a bit. Many older cars used a push button to start instead of a spring-loaded start position in the ignition switch. This push botton may be bad.

Battery Cables

Now let's look at the signs under the hood. The battery cables may be the source of the problem. Heavily corroded cables can make an otherwise charged and healthy battery seem to be discharged. Heavily corroded cables can be the cause of poor idling. High resistance in the system can also keep the battery from being fully recharged.

The corrosion that counts is not that which you can see. It is the corrosion between the battery post and clamp. This corrosion, like corrosion or oxidation on any electrical switches, is resistant to the flow of current. It can be so bad that even with a new fully charged battery barely enough current gets into the battery cables to light a dashboard bulb. This problem seems to show up a lot during rainy weather. The best way to solve it is to use a special reamer for battery posts and clamps on the battery and cables, then put the clamp on the post, tighten, finally cover the clamp and post with a little grease. I have had spray-on anti-corrosion stuff seep down between the clamp and post and create a problem just like corrosion.

If the battery clamp is heavily corroded, replace it. Or cut off the old cable, strip and clean the end, and put on one of those 50-cent replacement clamps. It's a lot cheaper but lasts about half as long. If you decide to do it and recycle the copper in your old cable, know that you now have two sources for bad corrosion connections: at the post clamp and at the cable clamp.

Sometimes the bad connection is at the other end: at the connection between cable and solenoid. Look for a wisp of smoke or feel the terminal gingerly to see if it's hot or even warm after starting or trying to start the car. If so, clean things up with steel wool and tighten down again. On the ground-strap side, the cable to the motor or the body can be dirty or corroded. If the ground cable goes to the body under the battery holder, as it does in some Ford Cortinas, look for trouble there sooner.

Incidentally, anything mounted under the battery runs a high risk of being damaged by battery acid. Strange examples I have encountered have been some frozen clutch-linkage parts in an Austin America, and a continuously malfunctioning factory-installed capacitive-discharge unit in a Mercedes 220. If the battery is overfilled or leaking, neutralize the area with a water-and-baking-soda solution; a tablespoon of soda to a glass of water.

VWs, especially the transporters, have trouble with battery and cable maintenance. In addition, the 6-volt VWs use a wire from the ignition switch to the solenoid of too small a gauge.

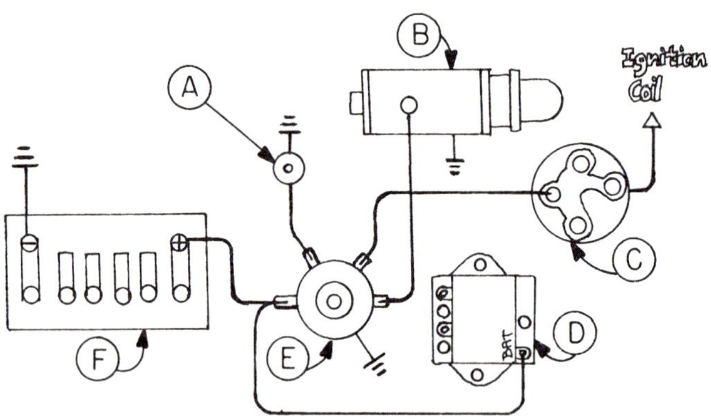

A Neutral Safety Switch (Automatic Trans. Only)
B Starter
C Ignition Switch
D Regulator
E Starter Solenoid
F Battery

Many people install another relay and have the ignition wire close a relay, which closes the solenoid relay, which closes the starter-motor circuit—wheels within wheels. The VW dealer carries this modification .

Battery

With a hydrometer you can measure the specific gravity in each cell of the battery. These readings have to be corrected for temperature. I find this method messy; besides, the hydrometer is made of glass, easily breakable. I use a Peerless battery checker. It measures the battery voltage with a load on it. Simply measuring the voltage of the battery is not sufficient because the voltage is not a direct measurement of battery capacity. Unfortunately, the Peerless checker will not identify a short in one cell in an otherwise healthy battery. A constantly high-charging rate indicated on an ammeter may be due to such a short. Either a hydrometer or replacement with a good battery will reveal this deficiency.

When the temperatures dip down to freezing and below, the battery loses its capacity but it merely needs to be warmed up to work properly. Hooking a trickle charger on the battery overnight will keep it warm. With the battery, the warmer it is the more capacity it has. With a motor (such as the starter motor), the warmer it is on the windings the more resistance it has, therefore the less efficient it is.

If the battery is in low charge, remove it and have it recharged. To remove a battery, remove the ground cable first. If the ground has been disconnected first, then you won't accidentally make a short circuit while you remove the hot cable that could fry the solid-state electronics, which are part of a car's electrical system these days.

If the battery won't take a charge, get a new one. In some areas "rebuilt" batteries are available. I would advise against these unless the car is known to be in excellent electrical shape. These batteries won't recover too well from heavy use. You generally get what you pay for.

Batteries lose charge slowly just sitting. If the battery is out of the car, put it on a piece of wood. It will discharge more rapidly, if put on the ground or on concrete.

Electrical-Leak Test

Anytime the battery cables are off the battery, especially because of recharging a dead battery, I do an electrical-leak check. Replace the hot clamp on the hot post of the battery. Make sure that all appliances are off, and disconnect the transistorized regulator, if there is one. Touch the ground cable to its post to see whether it sparks. If it does, you have a leak which may be a short. Refer to *Electrical Systems IV* to trouble shoot this problem. A dead battery could, of course, also be due to a generating system that fails to recharge the battery. Check this system as described in *Electrical Systems II.*

Once you have solved the problem or ascertained that no short exists, remove the ground cable. Then put on the hot cable and tighten, followed by the ground cable. Put a little chassis grease on the exposed metal parts of the clamp.

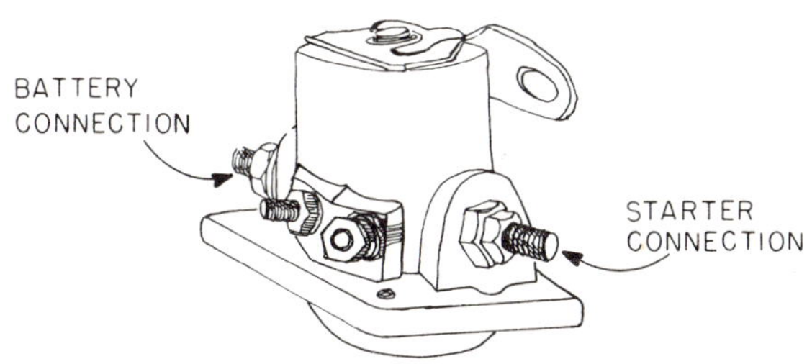

Starter Solenoid

If the battery and battery clamps are in order, and if there was no click when the ignition switch was turned, check the starter solenoid. The solenoid is mounted on the starter motor in most cars, the exceptions being some Fords and cars with Lucas electrical systems, where the solenoid is a separate unit mounted on the firewall. It is a heavy duty electrical switch that turns on and off the 100-to-300-amp current following through the heavy battery cable to the starter motor.

The solenoid is operated by the spring-loaded position of the ignition switch or by a push button on a few cars. To eliminate the ignition switch as the source of the problem, take a alligator clip lead and attach one side to the large hot battery cable. Touch the other end to the small-wire terminal of the solenoid. If there is more than one small-wire terminal, try both. If this is a Ford, it will be the S terminal. If this is a Ford with electronic ignition, disconnect the wires from the S terminal before trying this test. On older Lucas systems you can push a button on the starter solenoid to operate the switch mechanically and thereby check the starter and starter motor. I worked on a 1950 Studebaker in which the ignition wire terminal on the solenoid had to be "grounded" rather than "hot wired," as described above. If the starter motor now works or if the solenoid at least clicks, the problem is in the ignition switch or the wires running from it to the solenoid. If nothing happens when the solenoid is hot wired, it is fairly safe to say that it is bad and should be replaced.

Almost all solenoids may be replaced apart from the starter motor. Disconnect the ground strap from the battery first. The solenoids on the Delco type starter used on most cars also connect with an arm that moves the pinion gear to engage the flywheel ring gear. This is easily disconnected.

Starter Motor

If the battery, battery cables, ignition switch, and solenoid check out alright, we are left with the starter motor. There are two general types: the early Ford or Bendix type, and the General Motors or Delco type. Lucas is similar to the Bendix, while most other are similar to the Delco starter. Chrysler interposes a step-down gear between the motor and the pinion gear in some of its starters, which means that the motor runs faster giving a characteristic high-pitched sound. Otherwise it operates like a Delco starter.

Ford now uses a starter of the Delco type or another type that retains the feature of a separately mounted starter solenoid. In the latter type, a movable pole is closed when the starter motor is activated moving an arm that engages the pinion to the flywheel. This type has an overrunning clutch similar to the Delco type. It can be distinguished from the Bendix starter by the housing on the side that covers the movable pole device.

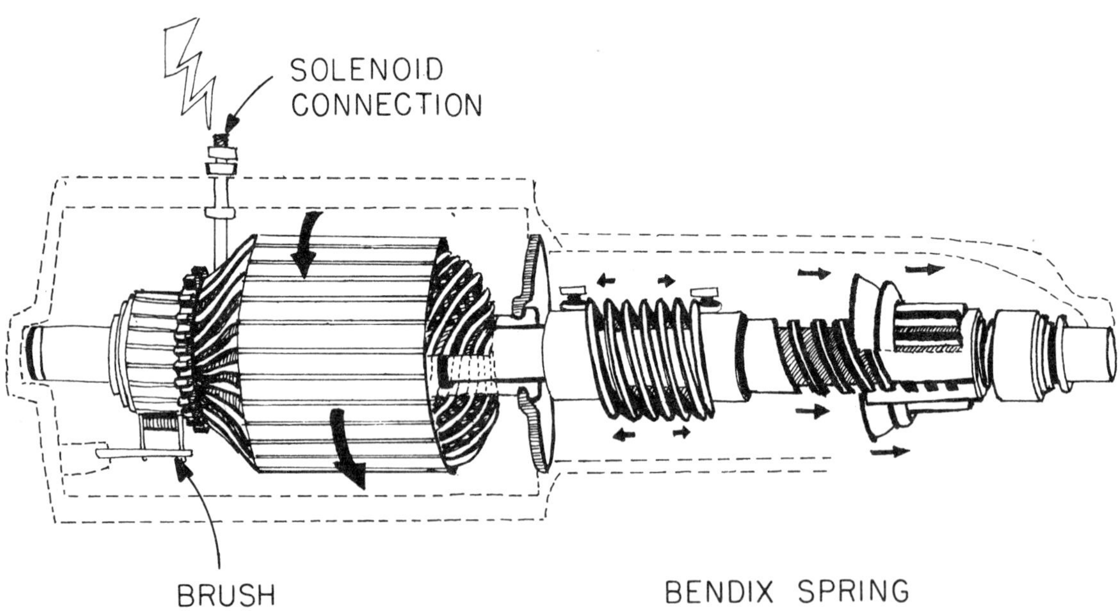

Bendix Starter

A Bendix-type starter may spin when the ring-gear teeth are knocked off. Since the engine tends to stop in just two or three spots, these few spots tend to receive all the punishment of starter-pinion engagement. Lucas starters, particularly, seem to punish the ring gear. (The pinion gear on the starter-motor shaft simply gets thrown into engagement with the ring gear of the flywheel to turn the engine over.) To distinguish between this problem and a broken Bendix spring, move the car while in gear or by some other means turn the engine a bit, then try the starter again. If it works now, you have some missing teeth on the ring gear. This trouble is fixed if the transmission, the clutch, and the flywheel are removed (see *Clutch*). (The pinion gear on the starter-motor shaft simply gets thrown into engagement with the ring gear of the flywheel to turn the engine over.)

Another characteristic problem with this type is that the Bendix spring breaks, which leads to the starter motor spinning but the engine not turning over. The Bendix spring can be replaced by a general mechanic if you can fine one. Also, the pinion gets stuck in the ring gear occasionally. You turn the key or push the button, and click–nothing. Actually a little more than nothing, because the pinion flies out and locks into the ring gear. But the next key turn is click–nothing. Some Lucas starters have a square-ended shaft (covered by a cap) opposite the pinion gear end. You can get a wrench on the shaft and turn it a little one way or another. This is usually enough to free the Bendix unit and you can try again. Otherwise you can put the vehicle in third gear if it has a standard transmission and push it back and forth. When the Bendix releases, you should hear a snap. Other methods are to strike the starter smartly with a hammer, or loosen the bolts holding the starter to the bellhousing. All of these troubles are the reason why the Delco-type starter is the dominant type on the market today.

When the engine starts, the pinion gear gets thrown out of engagement with the ring gear because the engine is spinning the pinion at high speed.

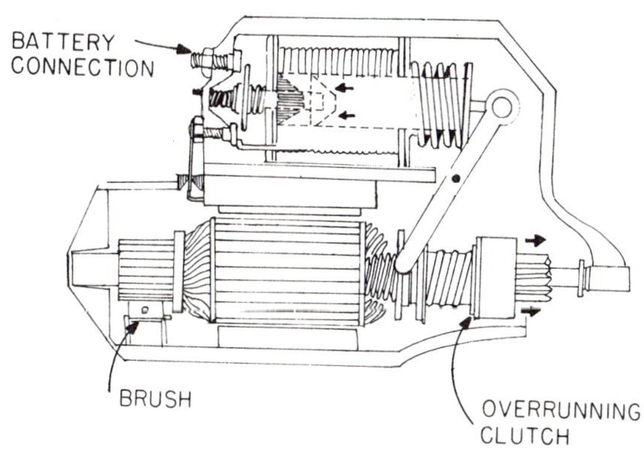

Delco Starter

The Delco type of starter has a more positive pinion to ring-gear engagement because of a shift lever connected to the solenoid. Therefore it doesn't have the problems of the Bendix spring breaking. Nor does it have much problem with the pinion sticking in the ring gear of the flywheel.

It does, however, have trouble with the overrunning clutch. This clutch protects the armature shaft from being spun too fast by the motor when it starts. Since the ratio between the ring gear and the pinion gear is about 15 to 1, if the engine is turning at 1,000 r.p.m. the starter armature speed would be 15,000 r.p.m. The overrunning clutch drives in one direction, but slips in the other. It slips when the engine starts to overdrive the pinion. The pinion returns to its unmeshed position when the driver relaxes the starter switch. When the overrunning clutch gets old, it tends to disengage before starting the engine.

Starters are generally very reliable. They don't get used continually like the generator or alternator, so they generally last longer. A replacement of the overrunning clutch of the Delco-type starter is certainly a repair worth doing by the general mechanic.

While the starter is out it should be checked for commutator and brush wear. If the commutator is grooved, or if the brushes are close to being flush with the top of the brush holder, consider rebuilding the starter.

I have encountered two other problems with the starter system. If the spring is broken in the ignition switch, so the key doesn't normally relax to the normal running position, the starter may be left inadvertantly in the start position; this sometimes causes the starter motor to overheat and burn up. And occasionally there will be a dead spot on the commutator; at times the car will start normally, but at other times it won't. These two problems require starter-motor replacement. In the first instance the ignition switch should also be replaced.

Another somewhat elusive problem is starter-motor overheating due to unfortunate placement of the motor near the exhaust manifold. The efficiency of an electric motor is inversely proportional to temperature. After a hot run on the freeway it may be difficult to restart until the starter cools down. Some cars are more prone to this problem than others. At the same time, the engine ring-to-cylinder resistance may be high if the oil has been thinned down, placing an even greater strain on the starter. Let the engine cool and don't pump an accelerator pedal, which will only flood the engine and increase ring-to-cylinder resistance.

Starter-System Test

To check integrity of the starter system while it is still in the car requires a voltmeter and an ammeter. Place the voltmeter-test leads on the battery posts, positive to positive, negative to negative. Place an induction ammeter over the battery cable that runs to the starter solenoid. Remove the high-tension cable from the center of the distributor cap and ground it, so the engine won't start. Close the starter switch for 15 to 20 seconds. Take an average reading from the ammeter.

The normal current reading for a 12-volt system would be from 75 to 200 amps (150 to 400 amps on a 6-volt system). If the battery reads 10 volts or more (5 volts or more on a 6-volt system), then the battery is in order. If both the battery voltage and the amperage are normal, then the starter motor, cables, switches, and battery are in good shape. If the voltage reading is low, the battery is at fault. If the amperage reading is normal or less than normal, and the starter is turning very slowly, then poor connections are indicated in the cables or the starter motor. If the amperage reading is high with the voltage normal, then an extra load is indicated on the starter motor. This may be due to shorts or other defects in the electric motor or a stiff engine.

Starter-Motor Replacement

Disconnect the ground strap from the battery. Then disconnect the wires running to the starter. Two bolts usually hold the starter in place. They frequently do double duty by also holding the bellhousing to the engine. Getting at the nut or bolt on many smaller cars is a blindman's job, and I honestly don't believe a mechanic with large hands can remove the starter on some Toyotas, and only with difficulty on a VW and Porche. On these cars the bottom nut is on a stud, which is not so difficult to reach, but the top connection, which can't be seen, is a nut and bolt. I use a small vise grip to hold the bolt on the starter-motor side of the firewall. The new VWs, and Porsches, use a special bolt which doesn't turn and therefore the vise grip isn't necessary.

In nearly all cars I exchange a faulty starter motor for a rebuilt one or have the faulty one rebuilt by my electric-motor rebuilding connection. There is no reason why you couldn't rebuild a starter using *Auto Engines and Electrical Systems* by Blanchard and Ritchen. It means using a lath and a few other specialized tools which a general mechanic wouldn't have, however, and unless the motor is exotic and therefore very expensive (an antique car or a Mercedes, for example) it's not worth it to compete with the specialists.

I don't recommend the "rebuilt" motors sold at discount houses. They generally repair only whatever caused the motor to fail but do not actually rebuild it. Then a year later you will find your "new" starter motor is acting up again.

For special problems or when a car is going to be laid up for a while for a complete overhaul, I have the starter motor overhauled in a local electric-motor rebuilding shop.

Chapter 9:

Electrical Systems III: Ignition System

There is no thrift in nature. There is rather waste. Human interests require thrift, selection, and preservation.

—William G. Sumner (1840-1910)

When the internal-combustion engine is being turned over by the starter motor, three things are necessary for the engine to start: compression, gas, and spark at the right time.

First, then, absence of compression sufficient to start the engine is unlikely although it may contribute to the difficulty of a hard-to-start engine. When the compression is low throughout, the engine spins over very easily. It is often difficult to accept low compression as the cause of the problem, so a compression test should be done if this is suspected. It would generally occur on a high-mileage vehicle with a lot of fast freeway miles on it; a salesman's car for example. Over-all low compression points to the need for engine overhaul.

Second, the absence of gas may be the cause of failure to start. It is much less likely than an ignition failure, but it can be quickly eliminated by the removal of the air cleaner and by goosing the accelerator linkage. A squirt of gas should be observed in the throat of the carburetor. If it cannot be observed because of the position of the carburetor, it can probably be smelled. If there is any doubt, a capful of gas into the mouth of the carburetor should be sufficient to start the engine for a short period if everything else is alright. Even then, an empty gas tank is more likely than a complete failure in the gas-delivery system. If the carburetors have just been rebuilt or if the gas tank has run dry, it will take some cranking of the engine to refill the fuel lines and fuel bowl. If fuel delivery does seem to be the problem, consult the chapters on that subject.

Spark Check

Third, if the engine gives no sign of trying to start, no coughing and sputtering or backfiring, then the ignition system is probably to blame. Pull the high-tension spark-plug cable off one of the spark plugs and stick a known good spark plug into the receptacle. Ground the threaded base of the plug with an alligator test lead. Crank the engine over and observe the spark or absence of it. This test is safe to do with electronic-ignition systems. If there is no spark, try a few other plug wires. If there is a spark in some but not others, the problem is in the rotor, the distributor cap, or the wires themselves (see *Tune-up*). If there is no spark at all, the problem is usually in the primary ignition system, which includes the points, condensor, coil, and ignition switch. Occasionally there is no spark because the rotor is not turning. If there is a spark at any plug, then the points, condensor, and all electronic gadgetry are functioning, although not necessarily in time (see *Tune-up for timing*). You also know that the coil is working, but not if it is up to snuff. Further checks are required to determine this (read on).

The next step in tracking down a no-spark condition is to use a test lamp and see if the battery or ignition wire to the coil is hot when the ignition switch is on. If it is, the test lamp will light up when touched to the ignition terminal and ground. If the lamp does not light, the coil is not getting current from the battery. Temporarily connect a test lead to the hot side of the battery and to the ignition side of the coil. The car should now start. The difficulty is centered on the ballast resistor, the ignition switch or the wires connecting the two from the battery to the coil.

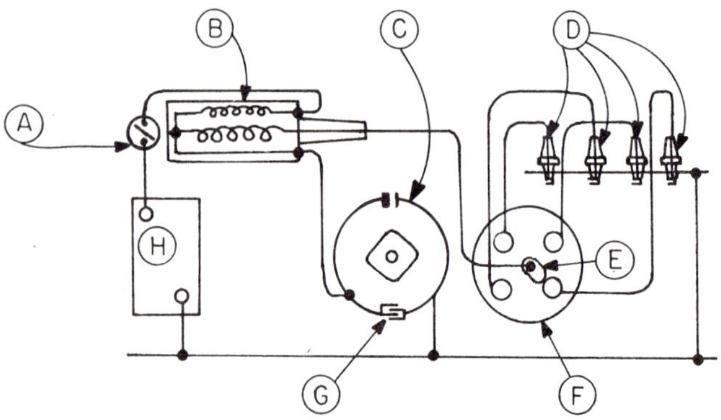

A Ignition Switch
B Coil
C Points
D Spark Plugs
E Rotor
F Distributor Cap
G Condenser
H Battery

Ballast Resistor and Ignition Switch

The connection from the ignition switch to the coil on many American automobiles is a special resistance wire and should be replaced by similar wire available at the dealer. Some cars use a ballast resistor mounted on the firewall. Most foreign cars have this resistance added inside the coil itself.

The purpose of this resistance is to reduce the voltage going to the coil from 11 volts to 9 volts. This reduction was necessary when 6-volt systems changed to 12-volt systems; the coil remained the same but a resistance was added in series to reduce the voltage. This change introduced another part which could potentially fail, the ballast resistor, but it also allowed for quicker starting by the following method. A second ignition wire was added, which ran from the starter solenoid directly to the coil, bypassing the ballast resistor or resistance wire. This means that during starting the coil receives the full battery voltage until the key is relaxed to its normal ignition position. This results in a hot spark for starting. If the coil would continue to receive full battery voltage it would overheat, have a short lifetime, and burn the points.

To test this resistor and switch, connect an alligator test lead from the distributor terminal (marked – or 1 or CB) to ground which is the same as closing the points so that the coil is charging. Then hook a voltmeter positive lead to the positive hot battery cable (in a negative ground system) and the other lead to the ignition terminal (marked +, 15, SW) of the coil, and turn on the ignition switch. The reading should be from 4.5 to 7.5 volts. When the key is switched to the start position, the voltage reading should drop to 1 volt or less. If this is a 6-volt system or if the resistance is built into the coil as it is with most foreign cars, the reading should be 1 volt or less all the time. (For further understanding of this voltage drop method of measuring resistance, see *Electrical Systems I.*) To determine whether a car has an external resistance, and bypass system, look at the starter solenoid. Generally if the car has more than one small wire running to the solenoid, it has a bypass. Cars with electronic fuel injection also have an extra solenoid connection.

If the voltage reading in the above test is zero, check for a shorted resistor. If the resistance is high, check for a defective resistor, loose or corroded connections, a bad ignition switch, or a bad neutral safety switch (in an automatic-transmission car). A few cars have a resistance wire that is in operation during both cranking and running. In this case a reading of more than one volt is normal at all times. The actual resistance, and therefore the voltage, changes with temperature.

Coil Checks

Coil problems usually show themselves by hard starting, missing or hard acceleration, and missing at top speed. The easiest way to test the coil in the absence of special test equipment is by replacement with a known good coil. Only one coil is necessary for temporarily checking any car. A 12-volt coil with no internal resistor will check all cars. This coil will supply too much voltage to the points for 6-volt systems and for 12-volt-system cars that use internal resistor coils. It will cause the points to burn. Therefore if the problem was solved by the replacement, get a proper replacement. Some electronic ignitions use a special coil.

Another method of checking the coil is to pull the center high-tension cable from the distributor cap and hold it with a clothes pin to check how far the spark jumps while you hold it close to the good ground. Either let somebody crank the engine, or turn on the ignition switch and carefully open and close the ignition points with a wooden stick. If the car has an electronic ignition, do the test on one of the spark plug high-tension cables and crank the engine. A healthy spark will jump a distance of 1/4" to 3/8" or more. The length of the spark is not a complete test of the coil, because the test does not measure spark duration, which will be affected by shorted coils.

You can test a coil for shorts if an ohmmeter is available. Disconnect the three wires to the coil. First measure the resistance between the small negative and positive terminals; it should be around 1 to 2 ohms. Next, connect a test lead to a screwdriver and insert the latter into the high-tension tower, and the other lead to either of the small terminals. The range here should be from 3,000 to 20,000 ohms. Check for manufacturer's specifications. Finally, attach the test lead to a small terminal and the other to the case of the coil. This reading should be infinite resistance. Any variation would indicate shorts in the coil; if it is still working, it won't be for long.

The coil in the General Motors system is mounted on the distributor cap but is otherwise similar to any other coil. To measure the resistance between the high-tension connection and the ignition connection of this coil, put one lead on the BAT (= battery) connection of the distributor cap and the other to the carbon brush which ordinarily touches the center of the rotor.

Electronic-Ignition Checks

The various manufacturers have mumbo-jumbo diagnostic routines for their electronic ignitions, but they all basically follow the same pattern. Check *Chilton's* for the particular routine, as it may be the most convenient method for doing a particular car. The Peerless model 465 Electronic Ignition Analyzer, which is designed for problem solving in this area also has instructions for each type of ignition. This instrument can also be used for conventional ignition and electronic fuel injection (see *Fuel Delivery III*). Nothing is really needed except a simple ohmmeter and a test lamp.

You do not actually test the electronics boxes in an electronic-ignition system any more than you do in the electronic fuel-injection system. What you do is eliminate everything else as a cause of the problem. When this is done properly you have to assume the electronics are malfunctioning.

The first check is for spark. Pull off a spark-plug cable and insert a good spark plug. Ground the threaded base to the engine with an alligator test lead, then crank the engine. If you get a spark, probably nothing is wrong with the electronics. The timing of the spark may be wrong or the spark-plug wires may be confused, of course, and the coil may be weak if one is used. Refer to the previous section for checking the coil.

If there is no spark, quickly check the ignition wire to the coil (the ignition switch on, car accessories off) with the test lamp. If it is hot, you can assume that current is getting to the ignition system. If no current is getting to the coil and a ballast resistor is visible on the firewall check each connector of the resistor. They should all be hot with the ignition on. Replace the ballast resistor if this is not the case.

Some of the electronics boxes require current to operate properly. In a Chrysler electronic ignition, disconnect the wiring harness from the electronics box and probe receptacles 1 and 2 with a test lamp, the alligator clip attached to a good ground and the ignition switch on. They should be hot.

The Ford solid-state ignition requires the disconnection of a three-prong and a four-prong connector from the electronics control box. Probe female connectors 4 and 6 with the ignition on and the test-lamp alligator lead attached to a good ground. They should both be hot. Connector 6 will not be as bright. With the probe still in connector 6, crank the engine. It should remain on and in fact get brighter.

In an American Motors breakless inductive-discharge ignition, disconnect the four-wire harness from the electronics box. Probe the yellow wire on the female side with a test lamp and the ignition on. It should be hot. In situations where connections were not hot when they should have been, move back through the wiring toward the battery with the test lamp until you reach a connection that is hot. Then replace bad wires or clean up cruddy connections.

Most electronic boxes are supposed to be grounded to the body of the car. To test whether they are, hook a continuity light to a good ground and the probe to the electronics box. The light should go on. If it doesn't, try loosening and tightening the mounting screws until the light does come on. Make sure your continuity light is working by touching the two leads together.

The final check is on the electronics module inside the distributor which replaces the points. In General Motors cars, separate the two connectors coming from the distributor and measure the resistance of the module with an ohmmeter. Connect one test lead to one distributor wire and the other to the other. The resistance should be from 400 to 1500 ohms. Then connect one of the test leads to the distributor body leaving the other lead alone. The reading here should be infinite resistance. If the readings are otherwise, replace the module.

The Chrysler system is measured in the same way. The module resistance here is normally from 150 to 900 ohms. The ground test should once again show infinite resistance. If it is less, the module is shorted to ground and should be replaced.

The Ford system has a three-prong connector. The two parallel prongs are measured as above. The resistance of the module should be from 400 to 1,000 ohms. The ground test should reveal at least 70,000 ohms. The third prong is measured by connecting one lead of the ohmmeter to it and the other lead to a good ground. If the reading is not zero resistance, then the module should be replaced, as indeed it should if it fails the previous tests.

The American Motors system operates on a different principle. Measure the resistance of the pickup module as described above, separating the connector from the distributor. The resistance between the two leads should normally measure from 1.6 to 2 ohms at 77 degrees. On the control-box side of the connector measure the voltage of the white wire, it should be from 5 to 6 volts. If it is not, and if you made the other ignition tests, replace the electronics box.

While you work inside the distributor, check to see that the control module is securely screwed down and close to the rotating vanes on the distributor shaft, but not actually touching. A clearance of .008" should be sufficient.

Rotor Not Turning

Occasionally there is no spark because the rotor isn't turning. The rotor won't turn if the gear at the bottom of the distributor shaft is stripped. This gear engages the camshaft. The distributor gear is softer than the camshaft gear so that if anything goes it will be the distributor gear. It is fairly easy to replace as long as you know how to restore the distributor to the proper timing, which is lost when the gear strips. Before you remove the distributor, mark the shoulder of the distributor where it abuts the engine block so that you can replace it in the same orientation. Also, before you remove the distributor, check to see if the valve train is moving when you turn the engine over.

Another reason why the rotor may not turn could be that the cam itself is not turning. To check this, remove the valve cover exposing the valve train or the cam in an overhead-cam engine. If the cam is turning, the valves should be going up and down as the engine is cranked; you can see it turn in an overhead-cam engine. If not, the timing chain is broken (it connects the crank with the cam) or the timing gear on the cam is stripped if there is direct gearing of the cam to the crank.

These failures usually happen during starting when a sudden lurch occurs from rest to movement. A stripped distributor gear may result from a stiff oil pump, if the engine has recently been overhauled. The timing chain breaks when it has too much slack resulting from extended use.

If the problem is a stripped distributor-shaft gear, while it is out try turning the oil pump with a long screwdriver. It should turn freely. If it turns with resistance, replace the oil pump.

Reinstalling the distributor requires turning the engine over to a known position, namely to the timing marks that indicate the firing point for the number-one cylinder. With the valve or cam cover (overhead cam) removed, observe the position of the exhaust valve when the timing marks have been aligned. If it is depressed rather than relaxed, you are on the exhaust stroke and need to turn the engine over 180 degrees. If the timing marks are aligned and both valves are relaxed (closed), the distributor can be installed.

Place the distributor in such a position that, when it is all the way down and engaged with the oil pump, the rotor is pointing to the number-one spark-plug wire on the distributor cap. Either use the marks you made on the distributor when you removed it or consult *Chilton's* to find the normal position of the distributor when it is in place. Use a long screwdriver to orient the oil-pump notch so that it will engage the bottom of the distributor. If, on firing up the engine, you get backfiring through the carburator or no response, then the distributor is in the wrong position.

Replacing the timing chain or timing gear requires stripping down the front of the engine. The radiator, fan belt, fan, timing cover, and probably the water pump have to be removed. If a timing chain or belt is affected, the cover may have screw in the most unlikely places, so before you bash it off with a hammer be sure that all screws are out. Be sure you get the timing of the cam and crank correctly when you install the new chain, belt, or gear (see *Engine Overhaul*). A broken timing chain often means one or two bent valves or pushrods because valves remain open when they should be closed when the piston comes up on its last stroke before the engine quits. Of course you can take a compression check after the chain is repaired, but it would be nice to know ahead of time. You have to remove the head to replace a bent valve so maybe you should just go ahead and do a valve job.

Chapter 10:

Electrical Systems IV: Charging System

Just as there are mechanics of genius who do splendid work with a few inferior instruments, there are also people who know how to use their experiences so widely that a so-called scholar can hardly rival them.

—George C. Lichtenberg (1742-1877)

Before the car is started the battery supplies all its electrical needs; therefore a faulty charging system has no bearing on a no-start or hard-to-start problem. In fact, once the car is running, you could drive fifty or more miles on a fully charged battery. At that point the battery would have lost its charge and the car would come sputtering to a stop. Normally once the car is started and idling, the charging system is not only supposed to run the numerous electrical appliances of the car but also recharge the battery. Hence we can check the charging system only when the car is running.

Red Light

In most cars today a red charging light is the only dashboard measure of problems in the charging system. It is supposed to come on when the ignition key is turned on and to go off as the engine starts. It merely indicates whether the current is flowing from the battery or from the generator (alternator). Sometimes the light will flicker on and off while the car is driven, indicating a short somewhere, probably in the generator (alternator), being made and broken. It may cure itself, but probably it will get worse. If the red light comes on bright with the ignition key turned, and then goes out with the engine running except for a slight glow, there is high resistance (dirty connections) somewhere between the regulator and the generator light, or the alternator "idiot light" diodes are defective. When the red light comes on and stays on, stop the car and look under the hood.

Fan Belt

Since the generator (alternator) is run by a fan belt that often also operates the water pump and the fan which cool the engine, it is wise to see if the belt is broken. A water-cooled engine can probably run at cruising speeds on the freeway (engine "floating") without overheating, but it would not last long on a hot day in stop-and-go traffic. Air-cooled engines, especially those mounted in the rear, are very vulnerable to overheating when the fan belt breaks. In the opinion of a reader of the manuscript who specializes in this area, a belt could never be too tight for an alternator. He recommended to try rotating the alternator fan by a finger. If it could be moved at all the belt was too loose. It's worth a try.

The fan belt may simply be loose and slipping. If it is, it will polish the pulley. One should be able to depress the belt around ¼"; shorter belts less, longer belts more. A belt too tight, however, will destroy the front bearing in the generator (alternator). I check the fan belt when I tune the car, rebuild the engine, or at any other suitable time. If it has cracks, I replace it. A fan belt stretches with use, particularly a long one, as in the Chrysler Slant Sixes. It may stretch beyond the range of adjustment and have to be replaced. It usually isn't the fan belt that is the cause of a red light. It usually is either the generator or regulator.

Since the regulator and generator (alternator) are hooked together in a feedback loop, you have to break the loop to determine which of the two is faulty. If you just want to operate on the probabilities, then the generator (alternator) is the best bet because it is the more mechanical member of the pair. In the simplest and commonest case the brushes have worn out.

Voltage-Output Test of Changing Systems

The first step in checking the charging system for a low charge or high charge—after the battery is examined for good health, the battery cables for good contact, and the fan belt for tightness—is to hook a voltmeter on the battery, positive to positive, negative to negative. Then the car should be run at a fast idle for 5 minutes to get warm. If the voltage reading is 7 to 7.6 volts for a 6-volt system or 13.8 to 14.8 for a 12-volt system, assume the voltage regulator and generator (alternator) are probably alright.

If the voltage is too high, the battery may need refilling; in the extreme the battery will be destroyed and the filaments of all bulbs in the car will burn up.

If the voltage reading is high, the regulator is probably at fault or the field wire could be shorted in the generator (alternator) bypassing the regulator. Of course, the generator (alternator) may have failed because of improper voltage regulation and so will not produce any voltage at this point. Therefore examine the commutator of the generator, (the slip rings of an alternator) and look for spots of silvery solder caused by overheating. Or make the mentioned test again when the new generator (alternator) is installed, to see whether the regulator is keeping the voltage in the proper range.

Even if the voltage reading is normal and even if the red light goes out when the engine is accelerating, it is not certain that the generator (alternator) is properly working. Also, if the voltage reading was low in the previous test, then either the regulator or the generator (alternator) may be at fault. For these reasons to test the current-producing capacity of the generator (alternator), the regulator must be isolated. I will pause here to contrast the two types of generator systems. Following that, an isolation test will be described.

Generator vs. Alternator Systems

Both the generator and the alternator (which is in fact a generator also) produce alternating current, but the generator only uses half of its cycle: the half which matches the polarity of the battery. The alternator rectifies the current, which enables it to use all its output. This is why one polarizes a generator, but why one does *not* polarize an alternator. The alternator, being more efficient, produces between 5 and 10 amps at idle. The generator does not produce enough current at normal idle to run the headlights without help from the battery. Under these circumstances the red charging light will come on.

Part of the generator (alternator) system is a

feedback loop that shunts some of the current produced by the generator (alternator) back to its own field coils. This produces a much stronger magnetic field than that created by a permanent magnet. The permanent magnet merely starts the current flowing in the circuit. Alternators have no permanent magnet, so the initial field-coil excitation comes from the regulator. In some alternator systems the red warning light is part of this circuit; if it is out the alternator doesn't show an output. Replacing the bulb fixes everything.

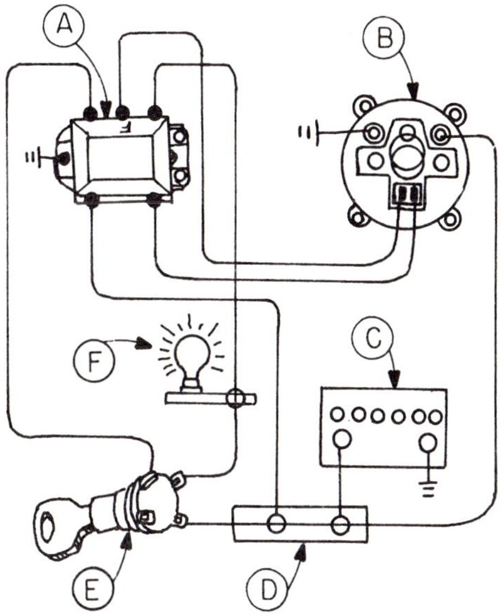

A Regulator
B Alternator
C Battery
D Junction Block
E Ignition Switch
F Charging Light

Regulator

With generators, the regulator has three functions. First, when the demand for current is low, the regulator controls the voltage output of the generator; otherwise excessive voltage levels which the generator is capable of producing might overheat the battery, blow fuses, burn up light bulbs, or fry solid-state components. Second, when the demand for current is high, the regulator controls the amperage output of the generator; otherwise excessive amperage levels might burn up the generator. Third, the regulator controls the flow of current to and from the battery. When the ignition key is turned off, the regulator breaks the circuit between the battery and the generator. If it did not, the battery would be discharged. This is done by the cutout relay.

With alternators, only one form of regulation is generally found: voltage regulation. The diodes in the alternator act as one-way valves to electrical current, eliminating the need for the cutout relay. If the red light remains on after the ignition key is removed, a diode is shorted to ground.

Current-Output Isolation Test

To determine whether the generator (alternator) or the regulator is to blame for a low charge or to check the current-producing capacity of the generator (alternator), an isolation test may be done. I have read much confusing information on this diagnostic procedure. The steps which I will outline are basic, I think, for general application.

With the ignition and all accessories *off*, disconnect the field wire from the generator (alternator). On an alternator it may be a double plug, with the field wire marked "F." If you are checking an alternator system and can find no field-wire terminal, then the regulator is solid state, integrated with the alternator housing. In this case no distinctions need to be made; remove the unit for repair.

Bypass the regulator with an alligator test lead so that the generator (alternator) will run unregulated. Since there are two ways to design the generator (alternator), there are two ways to bypass the regulator.

Ford, Lucas, and some Chrysler generator systems should be be passed by hooking the jumper cable on the field terminal (marked F) and the other end to the armature terminal (marked A or D). Most alternator systems also are checked in this manner. This is the method for internally grounded systems. Be careful when working around an alternator because the battery terminal is hot even with the ignition key off. Shorting this to ground could damage the alternator.

Most other generator systems on General Motors and some Chrysler cars, as well as Bosch generating systems, are bypassed if you hook the jumper cable on the field (marked F or DF) terminal and the other end to ground. This is the method for externally grounded systems. If you don't know which test to take, try both methods.

Next, place an induction-current indicator (that is the ammeter) over the armature-to-regulator cable, which is the largest wire connected to the generator (alternator), or over the battery-to-regulator cable, whichever is more convenient. Try to keep the meter more than 6 inches from the engine block or the generator (alternator). Also try to have 2 inches of straight cable on either side of the meter. Start the car being careful not to race the engine. Slowly increase the engine speed observing the meter. At fast idle (1,500 to 2,000 r.p.m.) the rated output for the generator (alternator) should be approached. Do not exceed that rating and don't race the engine. In a good unit the amperage should climb fast. If the amperage output is good and the voltage reading was low in the previous test, then the regulator is at fault. A low reading indicates a faulty generator (alternator); it should be removed for repair.

One unusual and rarely detected malfunction is a constant overcharge. It would not be indicated by the idiot red light on the dash and therefore shows up only if a test ammeter is placed next to the armature wire running from the generator (alternator). If the amperage level stays pinned to the right of the gauge (full charge) at all times except when the engine is not running fast enough, then the amperage controlling the points in the regulator are welded together or there is a short in the field-wire circuit. This short will cause the generator (alternator) to run hot and have a short lifetime. To distinguish between the possible sources of the problem, disconnect the field wire from the regulator. If the amperage drops to zero the problem is in the regulator. If the needle is still pinned to the right, the problem is the field-wire circuit.

On the other hand, if the charging gauge shows any indication of regulation, if the needle drops down at all with the engine at a fast idle, then suspect a short in one of the cells of the battery. A short can only be detected by a hydrometer or by replacement of the suspect battery with a good one.

Other indications of generator (alternator) malfunction are a loud whine usually in the front bearing, frequently a dead battery, a noisy radio, and dull headlights. Excessive sparking in a generator occurs when the brushes are worn down or the brush holders sticking. Very occasionally a recurring red-light problem and a failing regulator occurs every six months or so, possibly the result of a mismatch between the regulator and the generator (alternator). This leads to the sobering discovery that the manufacturers don't always have the best information on their own cars.

Generator and Alternator Replacement

Start any work requiring alternator replacement by disconnecting the ground strap of the battery first. If the ground strap is removed first, you do not risk shorts while removing the armature wire, which is hot even if the ignition is off.

Slacken the fan-belt adjustment and notice if the generator (alternator) pulley is lined up with the crank pulley. Occasionally you will find the wrong generator (alternator) on a car, especially if it is an old-timer. The wires to the generator (alternator) are usually self-explanatory. The connections are often unique, and the length, size, and shape of the wires are usually sufficient to remind you where they belong. The fattest wire is the armature wire (marked A, B, D, or D+).

On 6-volt VWs and Porsches which have the regulator mounted on the generator, I replace them both as a unit, because they come that way from my parts supplier. There are many repair places that routinely replace both units together whether they are mounted together or not. There are undoubtedly situations where one member of the pair causes damage to the other, but in my experience the generator (alternator) is usually the problem and should be replaced.

My general approach to repairing generators (alternators) is to trade them in for rebuilt versions. A good rebuilder will check the armature and field coils, replace the bearings, turn the commutator, undercut the mica, and replace the brushes in a generator. He will check the stator and rotor, bearings, brushes, and diodes in an alternator. The cut-rate parts outlets sell cut-rate rebuilding services. They generally only replace brushes, and in alternators also bad diodes. These adjustments you could do yourself for less money.

The commutator of the armature in the generator can be cleaned up with a piece of "00" sandpaper followed by a careful cleaning with trichloroethylene. A commutator with much wear is usually worn unevenly, which means that a new brush will ride the high parts until it seats in. To repair this, the commutator is turned on a lathe after which the mica is undercut. The mica is the insulating piece that separates each copper segment of the commutator. The turning and undercutting can be done for you at an electric motor-repair store. Brushes can be purchased at a good parts store. The brushes on some alternators are easily replaced, sometimes with the unit still in the car. Check *Chilton's*. If the alternator doesn't work after a polarity mix up or an inadvertent short, then probably a diode has burned up. Diodes are easily checked once they are exposed. Hook the continuity tester leads onto the diode lead and the diode case first one way, then reverse the leads. The light should shine one way but not the other.

It is useful to be on good terms with the manager of a local electric-motor rebuilding outfit. If the job doesn't have to be in by 9 A.M. and out by 5 P.M., you can get an excellent rebuild on just about any generator, alternator, or starter motor from such a place. You will also need a place to consult on special electrical problems.

The older electro-mechanical regulators can be adjusted and the points filed and cleaned. Copious information for setting them is available in the various repair-manual abridgments, particularly for European cars. But I have little trouble with this unit. You might, however, have to adjust the voltage-regulating relay. When the regulator cover is removed, either one, two or three relays will be seen. The voltage-controlling relay has the fine-wire winding. With a voltmeter attached to the battery posts, run the car at a fast idle until it is warm. The regulator is sensitive to temperature. A normal reading would be 7-7.6 volts for a 6-volt system and 13.8-14.8 volts for a 12-volt system. There may be an adjusting screw or simply an arm to bend which changes a spring tension. This will change the voltage level at which the regulator controls the generator (alternator). Keep the engine running, replace the cover, and watch the voltmeter. You may have to fiddle with it a bit to get it right. Take care not to short the relay to ground with the adjusting tool on the regulator cover. If you do, replace the regulator.

Polarizing the Generator

After a new generator or regulator is installed, the system has to be polarized. This need not, in fact should never, be done with alternators. With externally grounded generators (most General Motors, Chrysler, and Bosch), momentarily touch an alligator-test-lead between the battery and armature posts of the regulator. With internally grounded generators (Ford, Chrysler, Lucas) disconnect the field lead and momentarily touch it to the battery post of the regulator. Then reconnect it. This little piece of incantational magic is supposed to get the magnetic forces all facing in the right direction.

As an illustration of the craziness you can encounter in this area of car repair, here is one of my stranger alternator-regulator experiences.

It so happened that by my isolation test, the alternator on a three-year-old American Motors car failed. This explained why the red light stayed on all the time. I noticed the alternator had been rebuilt once already. This, coupled with the fact that it was a car for a small producer, prepared me for difficulties.

My concern was quickly justified when the parts man told me that this particular alternator was unavailable in rebuilt form. However, there was one of higher output available and often used on this car. I decided to kill two birds with one stone. Perhaps the smaller alternator was underpowered in the first place. I asked the parts man to check the regulator numbers for both the old type and the new type alternator. They were the same, so I figured I was in good shape.

I installed the new alternator. The red light now went out when the engine was started, and the voltage reading on the battery was 14 volts at idle. I figured that was that.

However, the next morning the customer came over to show me that the red light would not go out now when the ignition was turned off and the key removed. Suspecting a mismatch between the alternator and the regulator causing a short in the alternator diode, I called my alternator specialist. She said I did indeed need another regulator for the higher-output alternator. Next I asked a local dealer if he had one in stock. He confirmed that I needed another regulator, but he didn't have one. I called a dealer further down the line. He said that he did have the regulator I wanted but that it wasn't necessary, the old one should work: four opinions, two for, two against. At this point I was about to go mad when the customer with the troublesome car called back to report that the red light now acted perfectly normally. What can I say?

Chapter 11:

Cooling System

Heat flows of its own accord only from a hot to a cold object.

—Second Law of Thermodynamics

The explosion in the cylinder creates high temperatures. The engine's lubrication will burn at these high temperatures. When this happens the friction of the rings against the cylinder walls increases and they begin to cut each other. Shortly they will stick, causing the engine to freeze.

This area of the engine either has to be cooled by air circulation or water circulation. With air circulation, a fan is often used to increase the volume of air moving by the heads and cylinders. The surface of an air-cooled engine is finned to increase the surface area that can radiate heat. Breaking fins off an air-cooled cylinder reduces its capacity to rid itself of heat and a hot spot develops. The water-cooled engine can generally handle greater heat loads in a more compact space, hence water cooling is almost always used on larger engines.

The first requirement of the cooling system is *not* to operate when the engine is cold. Engine lubrication is inefficient until the temperature is in the range of 170 degrees to 220 degrees. Therefore we don't want cooling on a cold engine. One of the original water-cooling systems solved this problem nicely. The thermosyphon system circulated water through water jackets surrounding the cylinders to hoses connected to a radiator. (A radiator is a part that literally radiates heat.) The inlet hose was connected to the bottom of the radiator, the outlet to the top. The water just sat until it got hot in the cylinder area. Whereupon it rose (heat rises) through the top hose to the top of the radiator. The radiator had a fan to draw cooling air through it. As the water cooled, it settled to the bottom and thence to the inlet hose to the cylinders. Cooling therefore depended on heated water being present. Because circulation by this method was necessarily slow, a water pump was introduced.

The introduction of a pump to the cooling system meant that cooling would occur at all times, even when the engine was cold; therefore a thermostat was introduced.

Thermostat

The thermostat closes down the cooling system until the water in the engine heats up. On an air-cooled engine shutters are closed blocking the flow of air to the cylinders until they heat up. When they do, the thermostat expands and the shutters open. In a water-cooled engine a thermostat is put into the top hose to block water flow from the head into the radiator. If the thermostat is stuck open, the engine will take a long time to heat up. If it is stuck closed, the water in the block will come to a boil. If the vehicle is supplied with a sensitive temperature gauge, you can see when the thermostat is opening. The temperature needle moves steadily upward as the engine heats up, then makes a slight backward movement as the thermostat opens.

The thermostat may be checked by suspending it in a pan full of water so that it doesn't touch the pan. A thermometer placed in the pan will measure the temperature of the water as you heat it up on the stove. The thermostat should open within 5 degrees of the temperature marked on the thermostat (around 170 degrees).

Water Pump

The water pump is a self-contained unit usually placed on the front of the engine with the fan attached to it. It is run by a fan belt from the crankshaft pulley, and acts on the principle of a centrifugal air fan. The blades are mounted on a shaft that rides on a couple of bearings. In older cars these bearings could be lubricated. In newer cars they are "permanently lubricated." In other words, they don't last as long; or only about as long as a nonpermanently lubricated bearing that is never lubricated. Brass brushings are also often used instead of bearings.

Water pumps sometimes fail with a screaming noise that comes from the bearing without lubrication. A loose fan belt can also scream, therefore check it first. When a roller bearing is failing, it will scream first when it is cold but after a while become quieter. As it gets worse it will scream all the time. Sometimes water-pump failure is accompanied by knocking and rumbling sounds, and the fan will appear to be loose. Grab the fan and see if it wobbles. If it does, it is caused by a bad bearing as well.

Removal of the pump is straightforward. The fan belt should be loosened at the generator, inspected for cracking, and replaced if necessary. The smog pump and various other units may have to be removed to gain access to the fan and pump. On some overhead-cam engines the timing belt may be affected by pump removal. If this is the case be sure to wire the belt in place and don't turn the engine over while the unit is out, because you will lose the timing. Remove the fan then finally the pump. If you have a good arbor press and the bearings, you could rebuild the pump yourself. Otherwise the parts store will have one. A fan with a missing, bent, or loose blade can cause pump failure so check the fan and replace it if necessary. Overtightening the fan belt can also cause pump failure. On replacement make certain the mating surfaces are clean. Use a new gasket and some gasket cement.

Radiator

The radiator should be painted flat black, a color which radiates heat most efficiently. With age, the tubes in the core get clogged and the front gets plugged up with leaves and insects. This, of course, cuts down on the cooling efficiency. Fortunately radiators are easily rebuilt by local auto radiator rebuilding shops. Here is one case where the cost of repair is the same for a Mercedes as for a Chevy. The cost depends entirely on the amount of copper core used. The tanks on the top and the bottom are reused. In some cases you can have the old core rodded out instead of replaced. I have always rebuilt radiators and therefore have no experience with the rodding technique.

Radiators are soldered throughout. This means that if you attempt to make a spot solder on a leak in the core, you will create at least as many leaks as you stop. The cores are soldered when they are dipped into a solder bath so that everything is soldered at once. A leak in the tank or the tank seam might be possible to fix with the radiator in the car, if it is on the horizontal plane. The metal has to be clean and flux has to be used. Use a propane torch. Another method is to use Liquid Steel to plug the leak. Stop Leak, which is put in the cooling system, will work on small leaks and is worth a try. The type of Stop Leak with pellets, however, may cause blockages in the tubes in the core.

Leaks are more common in the modern automobile because the pressure in the system is getting higher. A higher pressure in the cooling system raises the boiling point of the liquid in the system. The pressure is controlled by the radiator cap, which is spring loaded. If the pressure gets too high, steam pushes past the seal and out through an overflow pipe. I have found occasionally that water loss was due entirely to a wrong radiator cap.

Leaks

Leaks can occur not only in the radiator, but also from hose connections or the hoses themselves, water valves, the water pump, past the head gasket into the oil system or into the cylinders, past the freeze plugs in the head or block, or from the heater radiator inside the car. Leaks are accompanied by a drop of pressure in the system. A special water-system checker is available, which replaces the radiator cap. It measures the pressure the system will hold.

A leak will reveal itself often by a rusty residue starting from the source of the leak. If the leak is in the radiator, the water and the rusty residue may be blown all over the engine compartment. Sometimes the leak occurs only when the engine is running.

The freeze plugs in the block, made of aluminum, are prone to leak and if one does, the others are suspect. Freeze plugs are supposed to protect the block from cracking if the coolant freezes. Freeze plugs are hard to see because they are often behind the exhaust manifold. When you replace one, it has to be cleaned. If the engine is in the car, it will probably be easier to knock the plug into the block and leave it there. It will do no harm. You can replace the old plug with another force-fit aluminum plug or you can get a rubber plug with a metal plate and a tightening nut, which is easier to install. The trick with the rubber plug is to have the sealing edge in the block clean and *dry*. If it is wet, the plug will pop out like a greased pig. Freeze plugs on foreign cars have metric sizes and may be impossible to find. I have solved some small freeze-plug leaks by enlarging the hole with a tapered reamer then screwing in an appropriate-sized self-tapping screw, putting red locktite on the thread and a copper washer under the head.

Blown head gaskets are discussed in the *Valve* chapter. They are usually identified by an oil/water mayonnaise residue inside the valve cover. If compression is leaking into the water system, you might see bubbling in the radiator. A rusty spark-plug electrode suggests water leaking into the cylinder. In, some aluminum-headed engines, overheating can warp the head and cause a blown head gasket.

The water valve, which controls water flow to the heater on many English cars has a tendency to leak at the mounting flange, because the valve is aluminum and tends to corrode or warp. It can be resurfaced.

Actually many parts of the water system are aluminum. Although it doesn't rust, it corrodes rapidly if salts are in the water. Therefore an additive that protects the aluminum should be put in with the water. Newer cars often require the use of a manufactured coolant that has this aluminum protection in it.

Radiator cleaner can be used to flush the water system. Warm up the engine, then add about a cup of oxalic-acid powder. Keep the engine running for a while and turn on the heater to get circulation throughout the system; drain the system and put in clean water; run the car to get warm again; drain, then use clean water again and run car again, until only clean water comes out.

Radiator hoses may look good from the outside and yet be badly deteriorated inside. If the hose is mushy, it is probably bad. When replacing hoses, take the old one to the parts store. There may be a preformed one available for your car, or perhaps a universal flex hose will do the trick. There is a type of heater hose (very expensive), in which you can tie knots without kinking, but most heater hose has to be nearly straight to avoid kinks. On foreign cars the preformed heater-hose replacements are even more expensive than the type mentioned above, so you might as well get the superflexible type hose at your regular parts outlet.

Fan

Even the lowly fan has undergone some interesting changes in the past few years. The changes seem to be an improvement on the whole. The limitation of the old fixed mechanical fan was that it didn't operate efficiently when you needed it most—at idle—and it did operate efficiently when you needed it least—cruising on the freeway.

The thermostatically controlled fan was the first improvement. This fan is uncoupled from the fan hub until the temperature gets high enough. Then, metal expansion locks the fan to the hub and it turns. This means that the engine doesn't waste energy turning it while it is warming up. To check the operation of this fan, simply observe whether it is turning when the engine is warm. Some of these fans you fix by replacement, others can be adjusted by use of a feeler gauge. The viscous damped fan, which slips as engine speed increases, is repaired by replacement.

The thermostatically controlled fan or the viscous damped fan are actually only a short step away from the old type, but the electrical fan is a major change. This fan does not depend on engine speed to determine its rate. When it goes on, it goes on at full speed. Also, it is controlled by a temperature sensor screwed into the cooling system. This means that the fan operates only when it is really needed. The drawback of this unit is that much more can go wrong. The fan can be checked by directly hot wiring it to battery voltage. The temperature sensor can be checked with a ohmmeter assuming you can find out the normal resistance values. Otherwise it can be checked by replacement once it has been determined that the fan is operating properly and that the wire connecting the two is in order. If you don't like the way the thermostat is running the fan, add a toggle switch under the dashboard, not to bypass the thermostat but to give you an additional method of turning on the fan.

With the fan we wave goodbye to the cooling chapter except to say that temperature is an important consideration in any part of the car, not just the cooling system, radiators, and water jackets. Overheating in the engine can cause overheating in the transmission. Automatic transmissions occasionally have their own cooling radiators and temperature gauges inside the car. The fluid should be changed in an automatic if it overheats. The generator has its own cooling fan because electric motors generally operate best on the cool side. And, of course, the brakes benefit from all the cooling they can get.

Chapter 12:

The Valves and the Blown Head Gasket

Give 'em the works. Fix it, there's always a way. Be hard boiled. The good die young.

—Carl Sandburg (1878-1967)

The valve job is a good introduction to major engine work before a total overhaul is done. In most types of cars (except VW) the engine can be left in the vehicle.

The head of the engine in the American automobile has evolved from being just the cover for the cylinders with spark plugs screwed in (the flat-head engine); to the carrier of the valves with all their attendant accessories; to the overhead-cam engine, which more or less splits the engine in half with respect to function and moving parts.

Actually the head has always been more than just the roof for the cylinder in the block. The shape of this roof and the dome of the piston define the combustion chamber. And here, after all, lies the heart (or perhaps the heat) of the matter.

The temperature in the combustion chamber has to be just right. If it is too low, the gas particles will condense on the cylinder wall, dissolving lubrication, and not igniting; choking the carburetor compensates for this fuel condensation by enriching the mixture; probably the most wear on rings and cylinders occurs while the car is warming up. If the temperature is too high, the fuel charge explodes before the spark plug ignites it and, in the extreme, can melt the piston.

The temperature should range from 180 de-

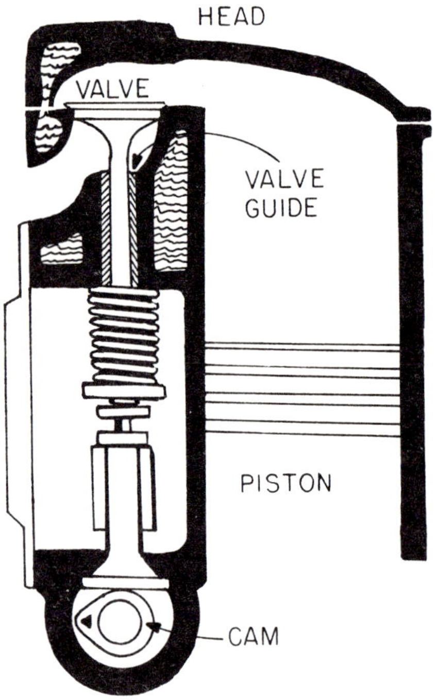

grees to 230 degrees, as reflected by water-jacket temperature or temeprature of the oil in the oil sump. Oil circulation contributes to engine cooling, which is why one sees large, finned aluminum oil sumps on high-performance cars. In the VW, oil contributes 35% to the overall cooling of the engine. I recommend that the oil-cooling capacity on the overworked 1600 VW bus engine is expanded by the addition of an extra oil-cooling radiator.

The actual temperature in the cylinders varies but is usually well above 1000 degrees. The exhaust valve is one of the hot spots, which is why it almost always has a .002" greater lash clearance than the intake valve on mechanically adjusted valves. This howling heat storm can cause erosion on the valves that looks like water erosion on bare land. Erosion will happen especially if the valves are adjusted too tight so that they can't close on the valve seat, or if a particle of carbon holds them open slightly. The valves are able to cool themselves off slightly when they rest on the valve seats which are being cooled by circulating water or air.

If the valves fail to seat and to maintain compression, performance will be poor and the idle rough; and low compression will be seen at the first step of a tune-up. These symptoms could also be due to a blown head gasket, broken rings, or worn cylinders and rings. To distinguish between ring wear and valves leaking, put a couple of squirts of 30-weight oil into the low cylinder and take the compression test again. The oil will temporarily seal the ring leakage. If there is no change in the compression, the valves are at fault. In practice, some improvement will be seen. In one case when I did this test, it pointed to leaking rings because compression went up, but it turned out that the valves were adjusted too tight. I took the test on the horizontal VW engine, which I suppose may make a difference in the accuracy of the test. Therefore a quick check of valve lash on one culinder would be wise.

On virgin factory engines, valve jobs are usually in order at 60,000 to 125,000 miles. The second time around, total engine overhaul is necessary. It's a bad idea to do just a valve job on an engine whose bottom end (crank *et al.*) has 100,000 to 150,000 miles or more on it. If one builds up the strength of compressive impulses on a tired rod bearing, the engine may throw a rod in 20,000 miles or less.

The head gasket seals between the block and the head. Compression can be lost when the seal fails between two adjacent cylinders. Water, and some oil (especially on overhead cams), passes between the head and the block; therefore, if the head gasket fails, one can also find gas bubbles in the radiator, water in the cylinders (the source for an invisible water loss), and water in the oil. Water and oil thrashed around by the crankshaft produce something looking like mayonnaise. Watch for this in the cap that screws into the valve cover where oil is added. Sometimes the engine can be seemingly frozen if one of the cylinders fills with water. Removing the spark plugs and cranking the engine will reveal a water fountain in this case.

Overheating problems on cars with aluminum heads often result in blown head gaskets and warped heads. The '68-70' model Saab overhead-cam aluminum head engines were, and are, notorious for this. The Vega aluminum block is also vulnerable to warpage due to overheating. It doesn't make much difference to the mechanic if the head gasket is blown or if a valve job is required. The head has to come off.

Cam in Block vs. Cam in Head

The cam in the head is brought into more direct contact with the parts it is designed to move, the valves, than the cam in the block. The head cam eliminates the pushrods required for the cam in the block. Fewer parts usually means more horsepower, other factors being equal. The practical differences for the mechanic are: when the overhead-cam head is removed, the basic timing between the crankshaft and the camshaft is tampered with, and the top of the head as well as the bottom is a machined surface. This means the valve job is more difficult and more expensive; usually at least 50% more than on a conventional head.

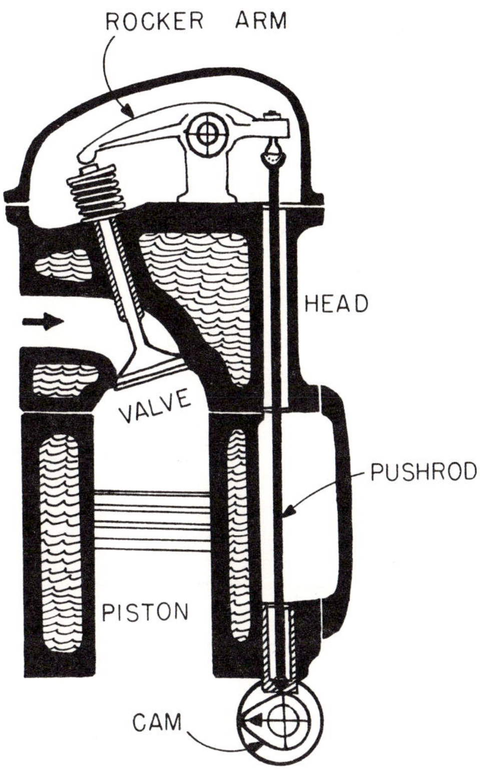

Push-rod-Engine Head Removal

The exhaust and intake manifolds have to be unbolted and pulled away from the head. Also, anything attached to it, like water hoses, water pump and coil, must be disconnected. The rocker-arm cover and the rocker-arm assembly must be removed. The push rods can be pulled out. The books say to keep the push rods in order. In practice, this isn't necessary; sort of like taking out tidal-wave insurance in Iowa.

Check the push rods for straightness by rolling them across a known flat surface like a piece of plate glass. If you notice any wobble, they should be replaced. If you find some badly twisted push rods and broken rocker arms, the driver may have overrevved the engine. The rocker-arm assembly is rarely touched on an engine overhaul, hence on an engine of unknown mileage one can check the amount of wear on the rocker arms. It will give a rough idea of whether the engine has 100,000 miles or less, or 250,000 miles or more.

Overhead Cam

With an overhead-cam head, mark everything but leave the cam in place if possible. Mark the timing chain and sprocket wheel so you get them back in exactly the same place. One tooth out and you're in trouble. Actually it is a good idea to tie the sprocket wheel and chain together, if you can do it and still remove the head. On the Vega you can do that, on the Datsun L-16 you can't. Mark the position of the rotor on the edge of the distributor; mark the

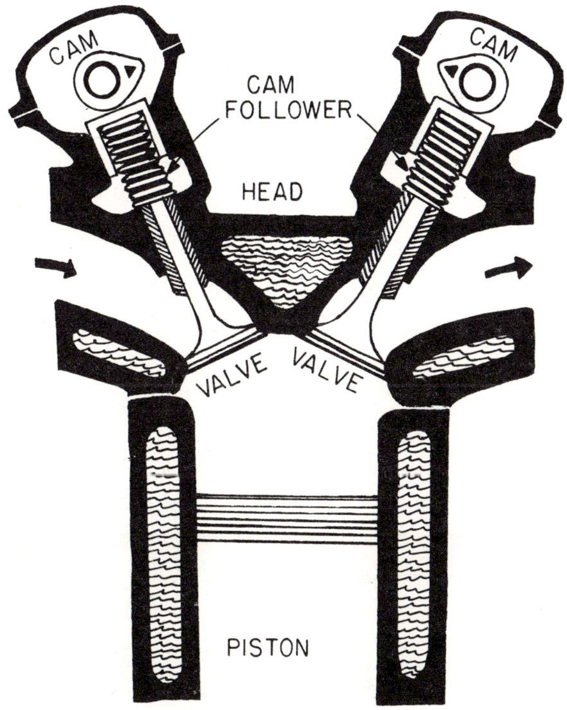

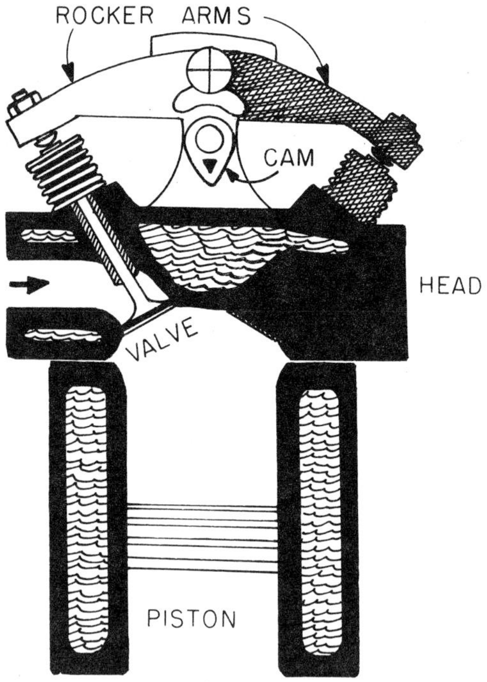

cam towers one through four, front to back; mark the position of the cam on the first cam tower and the position of the sprocket wheel on the end of the cam.

When you remove the sprocket wheel, keep tension on the chain so it won't fall off the crank sprocket wheel. To keep the top section of chain from slipping down below, you can jam a small board down into the space where the timing chain goes down into the bottom of the engine. On some overhead-cam engines such as the Opel 1900 and the Mazda 1800 (Ford Courier) the design of the timing-chain cover keeps the timing chain from coming off the sprockets of the lower gear. Consult a handbook or *Chilton's* for a particular engine to see what they say about it. If possible don't turn the engine at all now until the head and timing chain are reinstalled. It's a ticklish business. The timing chain is not simply the mechanical connection between the crank and cam. It establishes a set relationship between valve opening and closing, and piston position. If you lose the timing, the front of the engine has to be disassembled. The manufacturer's timing marks can be identified from a handbook or a factory manual. Lots of work! (For more information on cam/crank timing, see *Engine Overhaul II.)*

Actually, as a regular procedure I now disassemble the front of the engine; remove the radiator, then the fan, water pump, alternator and other belts, and whatever else is necessary so that I can get at the timing-chain tensioner. On many cars this is a spring-loaded nylon foot which presses out against the timing chain. This unit generally should be replaced if it is the sole method of tensioning the chain. When the head is back in place and the chain reinstalled, it can and should be retensioned. Careful people replace a timing chain with the first valve job, but I don't think it is necessary, if the tension can be adjusted. At the time of a complete engine overhaul at 150,000 miles or so it would be advisable.

The head should be removed when it is cool and the head bolts should be removed in a pattern opposite to that which will be used upon reinstallation. Don't set an overhead-cam engine on its head-gasket surface, because valves extend below the surface and may be damaged. Set the head on its side. The valves can be checked for seal with either a push rod or overhead-cam by turning the head upside down and positioning it with blocks so the combustion chambers are level. Then put gasoline in each combustion chamber. A leaky valve will let the gas drain out.

At the Machinist

This is our first encounter with the machinist, so let's dwell on the subject a bit. Along with the parts man, the machinist is a most valuable resource. It is important how you develop the first stage of your relationship. Machinists receive different instructions from different people as to how they should operate. From the junkyard down the road, which is selling "rebuilt" heads, the instructions may be to do the minimum, that is, just grind the valves.

In my relationship with my machinists I have tried to encourage a complete job, damn the cost. Of course, one runs into jobs where the cost of the repair is more than the car is worth, in which case I bring everything to a halt for a consultation with the person paying the bills.

It is amazing how helpful the parts men and machinists can be in a tight spot, and how difficult they can be if they have a grudge against you; so always pay the bills. In fact, until your credit is established, pay cash.

My dream is someday to find a parts-house/machine-shop combination that carries all the parts I need and can do all the machine work I require. I will never find it, but you can be sure that a machinist who has only the bare minimum in tools will sometimes downgrade the importance of (or even overlook) flaws that require tools he doesn't have.

Some of the things to be anticipated in a valve job and which increase the price are: valves in need of replacement, valve guides worn out, valve springs too weak, valve seats too thin, stripped spark-plug threads, cracked or warped heads, and metal corrosion.

Aluminum heads, which are the wave of the future, usually have more problems, especially warpage. Some Toyotas are notorious for cracks, which often make the heads worthless. VWs often have cracks around the valve seats, making them worthless. If an anticorrosion agent wasn't used, extensive corrosion can occur in aluminum heads, which heli-arc welding can repair but only at the risk of warping. Only a few machine shops can do this well. This procedure is only worth it if the cost of the new head is very high, as on a Mercedes. If it is an overhead cam, check what the thickness limitation is for the head because both sides may have to be resurfaced. I have successfully resurfaced both sides of a Datsun four-cylinder overhead-cam head. The local dealer junks both head and cam if there is warpage, at great expense to the customer. The key test is whether the cam will turn freely after being mounted on the resurfaced head, but before valve mechanisms are installed. A little bit out of line and the cam will jam. I have even shimmed a cam until it turned freely after being torqued in place. This is crazy but it can be done. It took me ten hours to do it, but this particular Mercedes head new would have cost the customer four times the cost I charged for labor.

Valve guides are a problem these days when the lead is being taken out of the gas. The lead acts as a dry lubricant to the valve stems. Bronze silicon guides are used frequently now because of their built-in lubrication. They don't have the structural strength of iron guides, but you can't have your cake and eat it too.

Worn guides can be knurled, which is a method of raising the metal to reduce clearance. One of my machinists calls this a cheap shot, but nevertheless the Ford Company, which buys up heads from the junkies to rebuild for their dealers, use knurling. Consult your machinist for his opinion.

Hydraulic lifters will probably last until engine-overhaul time, but check them for wear. There are two basic approaches to valve adjustment on this type of engine. The first approach is adjustment by turning down the adjustment nut a number of times; since the adjustment nut used here keeps it position by being slightly pinched and therefore tight (instead of a lock nut), I recommend that you replace the nuts unless you can be sure they are still tight enough. They are only available at the dealer. The second approach requires only that you torque down the valve-train nuts to a specific value. This is all well and good if it works. But if the engine comes in sounding like a noisy sewing machine, it's a good idea to keep your eyes open for the cause as you dismantle; especially watch whether the valve-train nuts are loose. Keep in mind that resurfacing the head, grinding the valve seats and edges, and using a thinner head gasket reduces valve clearance. The head gasket you get at the parts store will be different from the one the dealer carries, so you may have a choice.

The basic minimum for a valve job is usually one or two valves replaced, valves and seats ground, and exhaust guides replaced or knurled (they run hotter than intake guides and thus wear out sooner). On a blown head gasket it is advisable to resurface the head. A warped head may have caused the head gasket to blow. Order a valve-grind gasket set at the time you bring the head in for work. The machinist will probably do it anyway because there may be valve seals that have to be replaced by him.

While you're waiting for the head to come back from the machine shop, clean up the top of the block. Ideally all the carbon should be scraped off the top of the pistons, and all the gasket remains cleaned off the top of the block, exhaust and intake manifolds, water fittings, and so on. All particles should be removed from the cylinders and head-bolt screw holes. It's important that the head bolts screw in free and easy. With overhead cams there is a high-pressure oil hole that feeds up through the head, through a cam tower to the cam. See that it is clean. The Aston Martin has a small rubber doughnut that just fits in a hole in the head gasket at the oil connection. The Datsun L-16 has just a little copper gasket at the oil feed built into the head gasket. All these high-pressure oil connections are sensitive to leaks. The seals used in the VW oil-cooler column are similarly prone to leakage. If you are using gasket cement on the head gasket don't use too much near the oil feed because the cement may be squeezed into the oil hole blocking it.

If you are doing a valve job on a VW, you might consider removing the cylinder sleeves for honing or reboring, and reringing the pistons, or installing oversize pistons and rings, whatever is necessary. If you don't want to get into this, you can buy a clamp to hold the cylinders against the crankcase. Resurfacing the heads on a VW requires fly cutting which may be beyond your machinist's capacity. Go to a place that specializes in VW and Porsche. I always fly cut these heads if there is any evidence of blowby or black exhaust marks where the rim of the cylinder abuts the head. These marks reflect loss of compression.

Back Together Again

You've turned the corner and got your head back from the shop and a valve-grid gasket set they ordered for you. Everything else is clean and ready to go. If you have an overhead-cam head where the valve adjustment is done with shims, set the head on a bench, on two blocks of wood placed at each end. This will keep the head suspended off the bench and protect the valves from being damaged. The cam has to be torqued in place following the prescribed pattern and torque pressure. Now measure the valve lash. Write down the values and add or subtract the amount needed to achieve the correct valve lash for each valve. Remove the cam again and insert the appropriate shims. These are available in a range of sizes from the dealer. Get a selection of common sizes because you may have to go through the process several times before it is correct. Valve adjustment on other types will be done after head installation.

Head Gasket

The head gasket is the touchiest gasket in the engine and the most difficult to replace. The function of a gasket is to help two machined surfaces mate. If machining were done in other more precise and expensive ways, no gasket would be necessary. The gasket does its job by being compressible, which makes up for tiny irregularities in the mating surfaces. Ideally no gasket glue would be necessary but I oil paper gaskets, and glue cork gaskets into the oil pan and rocker-arm cover. Ideally one should always replace gaskets when parts have been separated, and this is the rule for head gaskets because the two parts aren't going to have the same surface or align exactly the same. One might stretch a point on the valve-cover gasket if it isn't completely squashed or brittle, and on a water-pump gasket, if it is in one piece, clean, and silicon-cemented.

Paint both sides of a steel- or copper-head gasket with your favorite miracle sealer. Either Red Permatex or K-and-W gasket glue with suspended copper will do. I've used quite a few and have only the recommendation that you use something made for that purpose. Rubber-and-asbestos gaskets need no sealer unless recommended. Place the gasket on the block right-side up. Usually the numbers are placed up but don't rely on it. Sometimes it says top or *oben,* and sometimes asymmetries show you how it fits when you line up all holes. If you are working on an older English car with a copper gasket, you might have to pound out carefully some bends with a soft-headed mallet before painting it with gasket cement.

Be as gentle as possible while lowering the head onto the head gasket on the block. Often alignment pegs or studs keep the gasket and block aligned. But on some cars, like the old Chevy Six, you will have to slide around a bit to get everything aligned to enable the head bolts to go through. On some cars, removing the hood is a must to allow you to climb into the engine compartment to do this operation.

Torquing the Head

Now tighten the head bolts. Follow the pattern given in the book, if one is available. If not, start from the center bolts and move in a criss-cross alternation from the inner bolts to the two ends. I make three passes. First I use roughly 1/3 of the final torque value, then 2/3 of the final value, then the final value. Aluminum heads usually take 40-60 lbs., iron heads 70-100 lbs.; but be sure you find a recommended value because if you overtighten a head bolt by even as small an amount as 10 lbs., you may snap the bolt. And that, brother, is a lot of trouble to correct. After the engine gets hot a few times I retorque the head in one pass. I usually undertorque the head bolts by 5 lbs. at the initial installation planning to pick it up at retorquing time. VW heads cannot be completely retorqued unless the engine is removed. Loose heads, a common VW problem, can sometimes be cured for all practical purposes by just retorquing the bottom row of bolts, which are accessible after the valve cover and rocker-arm assembly have been removed. Take cylinder compression before and after to measure the results. This process is against the rules but it does work on occasion nevertheless. The recommended values and torquing pattern can be found in *Chilton's*.

Now you just reverse the steps used to remove the head, using new exhaust- and intake-manifold gaskets, thermostat-housing gasket, and so on, which will be part of the valve-grind gasket set. I put Alumaseal in the radiator along with the water just as a precaution and to lubricate the water pump.

Adjusting Valves with Hydraulic Lifters

Adjusting valves with hydraulic lifters can be ticklish. Before you start the engine, tighten down the adjusting nut until all the clearance between the rocker arm and valve end is gone. This can be determined by rotating the push rod between the thumb and index finger while the adjustment nut is turned down. When the valve clearance is zero, you can no longer turn the push rod. Then continue to turn down the adjusting nut in the amount recommended for the particular engine. It may be from ½ to 2½ turns. This is just a rough approximation, however, because the best way to do it is with the engine running.

After making a rough approximation of the valve setting try to start the engine. Let the starter motor spin the engine until the oil pressure builds up and the valves start to open and close. Leave the valve covers off, so that you can watch the valves. If nothing is happening and you are sure of the ignition system and gas, pull out a spark plug and spin the engine again. You should hear a loud "whoosh" out of the spark-plug hole when your thumb is placed over the opening. If not, there's no compression, the valves are hanging open and are too tight. Take a full turn off every adjusting nut. Try again until you hear that whoosh.

When the engine is running, observe the precautions mentioned in the immediately following section, and complete the valve adjustment. Proceed from one valve to the next, loosening the adjustment nut until the valve "clacks" unless it already does. Then turn it back down slowly until the clack goes away. Then continue to turn it down the number of turns recommended for the particular engine.

To adjust the valves on an overhead cam while the engine is running, you have to construct a temporary cover out of tinfoil to put over the timing chain, so it won't throw oil all over the place. You may have to make a dam out of tinfoil to keep oil from spilling over the exhaust manifold on a V-8 pushrod engine as well, if you adjust the valves with the engine running.

Starting the Engine

When you first start the engine, look for water leaks and keep an eye on the temperature gauge. Take the valve-cover gasket off to see if each valve is getting oiled. You should see oil coming out of little holes in the rocker arms. Watch the oil-pressure gauge if there is one or watch the idiot light. Always, after breaking into the engine for repair, keep close tabs on the vital signs for the first five minutes of engine operation. Sometimes a minor oversight can cause an engine to be destroyed if the error is compounded by your failure to pay attention. There will be much smoke as oil burns off the exhaust manifold and the new gaskets get baked. This is normal.

Chapter 13:

Brakes, Axles, and Bearings

One submits and removes the axletrees from the wagon; in other words, contents himself with waiting. In this way energy accumulates for a vigorous advance later on.

—I Ching

Stirling Moss might be able to win a Grand Prix race without brakes, but for most of us a car without brakes is a useless car, out of control. It is essential to learn to "read" the brake pedal.

"Reading" the Brake Pedal

There should be a small free space in the first part of the pedal movement. Here you are working against the brake-pedal return spring. The function of this space is to insure complete disengagement of the pedal linkage from the master cylinder. The space stays fairly constant and won't be affected by adjusting the brakes. It is changed by adjusting the throw of the pedal linkage. Only in older cars does this cause trouble. The space can be very small but it must be there.

Next in the pedal movement, the pedal linkage pushes the master-cylinder piston in. You often have to do this by hand to feel and hear the two states.

Then the pedal should hit a solid wall where the brake shoes hit the drums or rotors. On new tight brakes it may take almost no pedal movement to reach this point. Springiness in the pedal may be due to the flexing of either the brake drum or a vacuum-assist unit, or both. Nothing needs to be done about it. On newly adjusted brakes the total pedal movement should be around an inch. As the brakes wear down this distance increases, unless the brakes are self-adjusting in which case the pedal height theoretically stays the same. On a mixed disk-and-drum system this distance increases, but more slowly; and on a pure disk-brake system the pedal remains high, because disk brakes are self-adjusting.

If, after you hit greater resistance, the pedal continues to move down, and you can raise the pedal height by pumping the pedal; if, in other words, you have a spongy pedal, there is air in the hydraulic lines. Probably the rubber seals of the wheel cylinders leak. Brake fluid is leaking

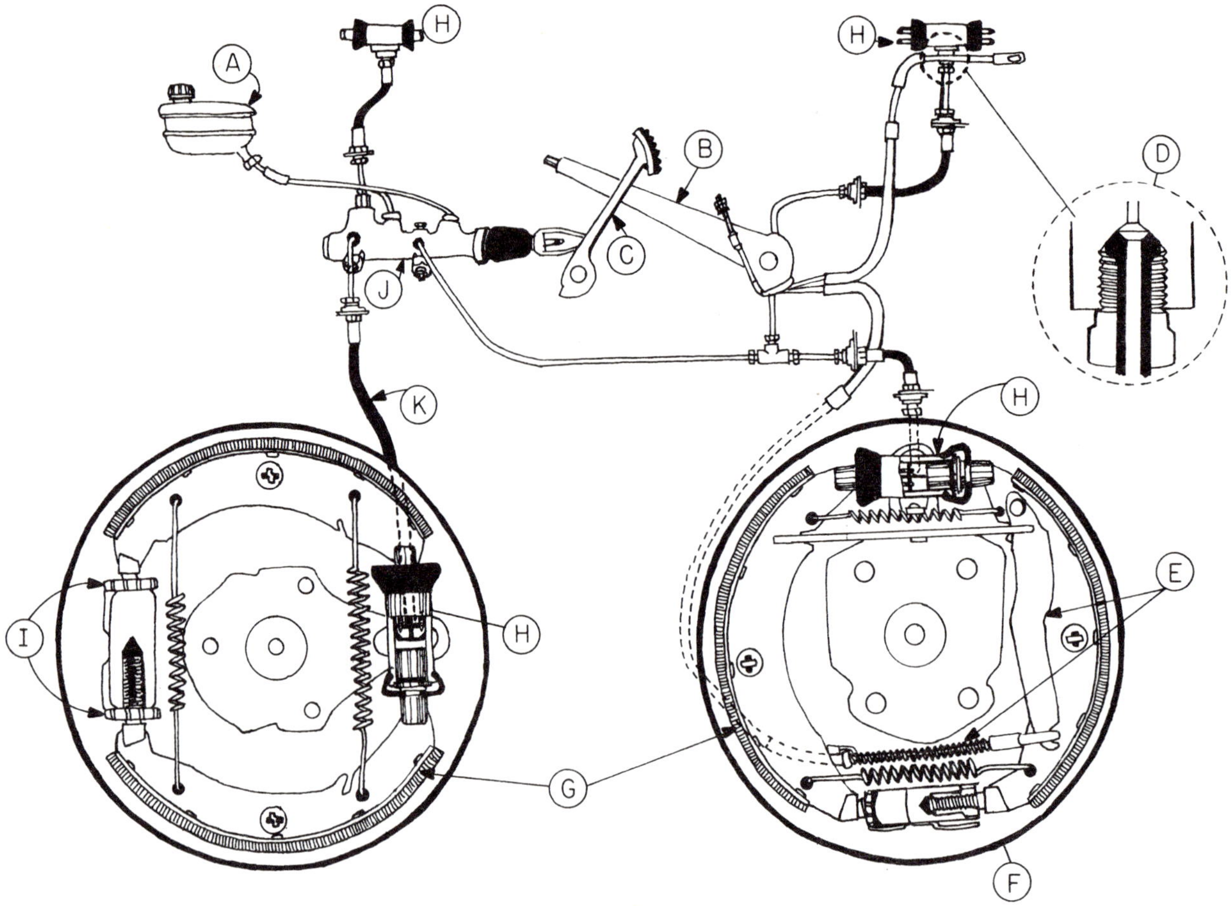

A Brake Fluid Reservoir
B Hand Brake
C Brake Pedal
D Double Flare Fitting
E Hand Brake Linkage
F Backing Plate
G Brake Shoes
H Wheel Cylinders
I Brake Adjustors
J Master Cylinder
K Flexible Brake Hose

out, air is leaking in. The brake-fluid reservoir will sustain this loss until it runs out of fluid. At that point, the brake pedal usually goes to the floor boards. Sometimes air in the hydraulic system will cause the pedal to press back out at you as you press on it, when the air in the line is first compressed and then expands as it gets hot.

If the pedal pushes slowly to the floor but doesn't pump up, and if its movement can be slowed by pushing harder, then the rubber seals in the master cylinder are faulty. The piston is pushing through the fluid rather than pushing fluid in front of it out to the wheel cylinders. In this case fluid will not be lost from the reservoir nor will there be any apparent fluid leaks on the brake backing plate (drum brakes) behind any of the wheels.

The leaking of brake fluid at the wheel cylinders in drum brakes often accompanies the grabbing of brakes. If one of the front brakes grabs, it will pull the steering to the left or right. This is very dangerous. If a rear brake is grabbing, the steering is not so seriously affected. If the steering pulls to the left, then usually it means that the left front-wheel cylinder has leaked on the brake shoes. For control in braking we ideally want the tires to maintain adhesion to the road, but the brake shoes to slip in their engagement with the brake drum.

The shoes should not slip too much, however, which can happen if the drum and the shoes become glazed and polished. If this happens (as it rarely does), the steering will pull toward the side operating normally. This points up another important need in the healthy brake system; namely that both sides are braking to the same degree. Another rather rare cause for pulling on one side is the jamming of a wheel cylinder, which makes that side effectively brakeless.

A visible leak of fluid on the rear-brake backing plate can be either brake fluid or 90-weight oil leaking from the differential. The two fluids can be distinguished by smell. Brake fluid often smells somewhat like alcohol whereas differential oil smells sulphurous.

Another sign of a faulty brake system is pulsing as you apply the brakes. It usually happens when the drum is out of round or the disc off-center. Have the machinist turn the drum or disc for you.

Wheel Sounds

Now we come to sounds that seem to originate in the wheels. Low- or medium-frequency rumbles that come from the wheel area, with the car rolling, engine off, transmission in neutral, brakes off, are caused by axle bearings. The sound may occur first on corners. In VWs, the rear-axle constant-velocity joints, or any constant-velocity joints for that matter, make a similar sound.

Sometimes a lug nut will be loose inside a wheel cover or the wheel cover itself could be loose. A clicking noise, especially around corners, could be a crack in the wheel between the lug nuts.

Brake sounds from the area of the wheels are usually higher pitched. Some brake shoes and pads squeak when they are new; this is alright, even if annoying. Sometimes squeaking brake shoes or pads can be quieted by wire brushing, then cleaning with trichloroethylene.

Brake shoes worn out to the metal make a sandpapery, grating sound. Then there are the sounds of broken springs and other brake parts inside the drum. Localize the sound, then remove the wheel and drum for a visual inspection.

Brake Repair

Front-brake linings last from 30,000 to 70,000 miles for either drum or disk brakes. The front brakes, which take 60 to 70% of the forward braking, wear out about twice as fast as the rear brakes. The hydraulic system seems to hold up well to the first 100,000 miles with some notable exceptions. I have had to replace all wheel cylinders on some Toyotas with 50,000 miles because they had jammed. It seems to be a Toyota characteristic.

Most brake work need only attend to the front shoes or pads the first time around. The second time perhaps shoes and pads all around, and attention to wheel cylinders. Older cars may need anything including the backing plates, drum, and springs.

Front-Wheel Bearings

Since wheel bearings are exposed in the process of repair on drum brakes and, in anything but simple pad replacement, on disk brakes. I have included them in this section on brakes. When the brake-drum rotor is removed the wheel bearings can be inspected.

Pop off the wheel cover, loosen the lug nuts half a turn, then jack up the front of the car. Remove the lug nuts and the tires. Some cars have left-hand-threaded lug nuts or left-hand-threaded axle nuts on the left side of the car. This is to counteract the tendency of right-hand-threaded nuts to loosen. If you see an "L" stamped on the nut, or copper-colored nuts, loosen by turning clockwise. When I remove parts of the wheel, from the lug nuts to the brake-adjustor parts, I keep them together with the wheel from which they came because there is often a difference between left side and right side.

With the tire off, the dust cap can be taken from the hub of the drum or rotor. Pull it off with big arc-joint pliers, then remove any cotter pins, castellated nuts, lock nuts, and washers. Bump the edge of the drum or rotor with the heel of your hand, and the bearing should pop out. If the dust cap was missing or the grease thin and black, expect to clean up the bearings for inspection at the very least. Rubbing a little of the bearing grease into a rag will reveal any metal flakes which might come from a disintegrating bearing.

Clean up the bearings in solvent if they look suspicious or if bearing sounds were reported. They should be bright silver, shining, and unpitted. If they are discolored, they have been overheating and the temper is out of the steel. Replace distempered bearings. The outer bearing on the axle is the one that usually goes bad.

The inner bearing is removed after the drum or rotor is taken off. To get the rotor off, the caliper has to be unbolted and removed. There may be shims, so notice where they came from and put them back the same way. Don't let the caliper hang by its flexible brake hose; hang it with a piece of coat hanger or rest it on top of the suspension.

Behind the bearing in either drum or rotor is a greased seal whose purpose is to keep the grease in the axle and the asbestos dust away from the brake. The inner bearing and seal can be knocked out with a large punch from the other side of the hub. Unless you are careful during removal and the seal is fairly new, it should be replaced.

Front-Bearing Races

If any bearings have to be replaced, the bearing race should be replaced as well. The race is the circular surface on which the bearings run. Damage to the bearings will be reflected on the race. It is driven out of the hub, where it is force fit, with a punch through the other side of the hub. Place the punch through a cutout in the shoulder against which the race abuts. The new race is knocked in with a hammer until it is level with the edge of the hub. From there on it is driven in with a hammer and punch or a socket slightly smaller than the race. Be sure you are not putting it in backward. For the completion of this job refer to the end of the chapter where the installation of the drum or rotor is discussed.

Rear-Drum, Seal, and Bearing Removal

The rear drums, instead of turning around an axle with the aid of tapered bearings, are fixed to an axle that has roller bearings supporting it in an axle housing. There are two types. In the first type the end of the axle is splined and the hub of the drum slides on the spline. It may be a straight spline (in the VW), which is held on by a single center nut or it is a tapered spline (in a Rambler or a Volvo). The tapered drums will have to be pulled off with a wheel puller. Don't get a cheap puller because you may have to go at it with a sledge hammer. The cheap pullers aren't hardened even though they may look good. The taper makes a rather tight fit so don't overtighten the center nut when you replace the drum.

The second type has a flange on the end of the axle, which has the wheel studs mounted on it. Chevies and many foreign cars have this type of drum. This means that with the wheel off and the handbrake off, the drum should come off. It is often held onto the flange by one or two recessed Phillips-head screws. These should be removed. If the car has been through salt water, the drum may be well fixed to the flange. A good soaking with WD-40 will help, as well as prying at the holes in the drum with a thin-bladed screw driver. The older Chevies had a paper gasket between the drum and flange to prevent the drum and the flange from rusting together.

If the drum is heavily grooved, the brake adjustment may sometimes have to be reduced (not tightened, but *reduced*) before the drum will come off. This is to allow the shoe to get over a ridge that has been cut into the drum.

With the rear drum off, notice if there is a fluid leak that does not originate from the wheel cylinder. This may be 90-weight oil from the differential. Leaking can occur when the differential is overfilled or the seal at the end of the axle housing is faulty.

To replace the rear seal or the rear bearing, either the backing plate has to be removed or the axle. If the seal happens to be mounted in the backing plate (uncommon), the screws holding the backing plate to the axle housing have to be removed, and the brake lines disconnected. The seal can now be driven out and a new one installed.

The more common type has a seal and bearing mounted in the end of the axle housing. Sometimes the bearing and seal are one and the same. Generally you reach them by removing the backing-plate screws, which also hold a retaining plate. With the retaining plate removed, the axle, bearing, and seal can be extracted. Leave the backing plate loosely hanging in place.

If there is no retaining plate holding the axle and its bearing in place, there is a C-clip at the other end of the axle in the differential. Exposing this clip necessitates the removal of the differential cover plate, a locking bolt, and cross shaft, so that the axle can be pushed in towards the center of the axle to remove the C-clip. Then the axle can be pulled out.

To get at the rear seal or bearing with a flange-type axle, first loosen the retaining plate bolts. Then pry on opposite sides of the flange with two claw hammers at the same time. If this doesn't work, you can buy or rent a special slide hammer to knock out this axle.

The rear-axle roller bearing may either be a press fit on the axle or into the end of the axle housing. The bearing can be removed from the axle by grabbing the shaft in the middle with a rag and banging the end opposite the bearing smartly on a wooden block. If this doesn't do it, take it to your machinist. Bearings that are press fit into the axle housing can be removed with a special puller or by locking two small vise grips opposite one another on the outside race. Then pry carefully with screw drivers on both sides.

Putting things together is usually simpler than taking them apart in this case. Be sure to clean everything including the drain hole through the backing plate, if there is one. This drainhole is supposed to allow leaking differential fluid to vent to the outside without getting on the shoes. Sometimes it works. If you can't get the paper gasket, which may come with this assembly, use silicone sealer.

Let's now get back to the brakes themselves.

Drum Brakes vs. Disc Brakes

The direction in brake design is toward the disc brake and away from the drum brake. This is an improvement. Still, an older Porsche or Mercedes drum brake in good shape is better than a new VW disc brake. Disc brakes are more efficient, have fewer moving parts, and pad replacement is exceedingly simple. If disc brakes are used in the rear wheels, some special arrangement has been made for the mechanically linked handbrake—usually a mini drum-brake setup in the rear-wheel rotor which, in my experience, often doesn't work very well. These drum brake shoes should never wear out unless the handbrake is left on by mistake while the car is driven. Some of the practical differences between the two hydraulic systems are:

Since the brake shoes have to be retracted away from drums by springs when no pressure is applied to the brake pedal, the wheel cylinders are always pulled back in. This means that unless the system leaks, the level of brake fluid in the reservoir stays the same. With the disc brakes the caliper pistons gradually extend as the brake pad wears off, and since no retraction of the pads occurs the level of fluid slowly drops. This is normal. If you add brake fluid, it will overflow the reservoir when the brake pads are replaced and the pistons retracted. If the cap on the reservoir is in place, you might have difficulty retracting the pistons. Since brake pads are constantly lying right on the rotor, they are automatically adjusting without the need for special gismos requried for drum brakes. The brake pads lying on the rotor also account for the slight scraping sound you hear, if the wheel is spun around while it is off the ground.

The other main difference in the hydraulics of the two systems is a flowback valve in the master cylinder for the drum-brake system which doesn't exist in the disc-brake system. It may be situated where the brake-line fittings comes to the master cylinder or at the end of the piston in the master cylinder. This valve keeps slight pressure in the system after the brake pedal is released. This keeps the seals expanded against the wheel cylinder walls and prevents fluid loss.

Remove the drums and make the following checks: Look at the thickness of the asbestos pad on the shoe at its most-worn spot. If the thinnest part is 1/16" or less, replace the shoes. If the shoes are not worn evenly, the anchor pin is improperly positioned. Unfortunately this fault is not adjustable on contemporary automobiles. If the shoes are worn down to metal, check the drums for deep scoring. If the surface is uneven, the drum will have to be taken to the machinist for turning (resurfacing). Some American cars with drum brakes and automatic adjustors have horrendous drum damage, requiring complete drum replacement for two or three times the cost of turning the drum. Some people remove the automatic adjusting linkage and treat the brakes as manually adjusting brakes. The bottom spring has to be replaced, if the automatic adjusting linkage is removed. Ask the parts man. I would recommend replacing all the springs and leaving the automatic adjusting linkage intact as a first attempt to solve the problem.

When you dismantle the brakes to replace the shoes or rebuild the wheel cylinder, first remove the hold-down spring or slips. The Bendix-type hold-down springs which must be pushed in and twisted are easily removed with a special tool made for that purpose. On some cars removing and replacing these springs is easy; on others very difficult. The retraction springs come off next. There is a special tool for this, but I never learned how to use it because I have been so successful with a screwdriver and a pair of parrot-beaked pliers. I made parrot-beaked pliers by modifying a standard set of Craftsman pliers with a grinding wheel.

The retraction springs are color coded. This is helpful if you have a book to explain the code. I remove the various parts and place them in a rough arrangement on the ground which resembles the way they were on the backing plate. You can leave one side assembled for reference. Never take all the parts off the wheels and throw them together into a cleaning can. When you take off the shoes, notice whether one of the linings is shorter than the other and where they go. Now clean the backing plate and the other parts of the brake with trichloroethylene. Don't breathe the dust from the brake.

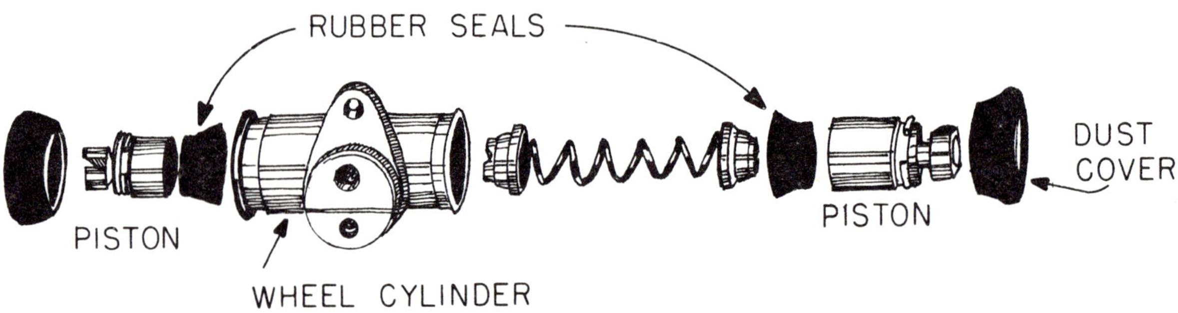

Wheel Cylinder

When a brake drum is removed, the wheel cylinders should be checked. If the inside of the drum and the brakes are wet, you know the cylinders are leaking. In the past I would not do anything to the wheel cylinders if I saw no leakage when I peeled back the rubber dust boot on each end of the cylinder. This has not always worked as a test (especially on some Chrysler cars). Some books mention a rule of thumb: measuring this clearance between the cylinder and the piston. If it is .004" or less the unit is supposed to be acceptable. I have found that this is not always true.

Therefore we have no flawless test to see if the wheel cylinders will leak if left alone or even if they are rebuilt. To be absolutely sure, then, you have to put new wheel cylinders on if you don't mind the expense. If you can be satisfied with 95% certainty, rebuild all wheel cylinders, rejecting those in which the bore is etched, scored, pitted, or in which the cylinder was loose on the backing plate.

If you plan to rebuild the wheel cylinder, another can of worms is opened: whether or not to hone the cylinders. It is inadvisable to hone any aluminum cylinder, but I advise against honing in general because the piston won't fit as well after honing. Most wheel-cylinder or, for that matter, master-cylinder failure seems to be due to the neoprene or rubber (Girling brake systems) seals becoming less pliant and less flared. With the aid of a kit that contains new seals learn to recognize whether the seals are good or not. This knowledge will help when you get a leaky cylinder or a recently rebuilt brake system. If the seal looks good and, especially, if it's an older vehicle, then a kit has probably been tried and has failed. Replace the whole unit.

If I am rebuilding a wheel cylinder I clean out the bore with steel-wool then follow with brake fluid and a clean rag to clean out any steel-wool particles. Don't spill any solvent, gas, oil, or grease on the neoprene or rubber parts of this unit. Use brake fluid to prelubricate the units as they go back together. Petroleum products cause the seals to deteriorate. If you are working on a Girling brake system, be sure to use Castrol or Girling brake fluid, because Girling uses rubber instead of neoprene parts and rubber needs a special mineral-based fluid. If you found rust in the wheel cylinder, flush the complete hydraulic system with new break fluid to get out the water.

The seals go in each end of the bore, concave sides facing each other. The flat sides abut the pistons. The spring in the middle positions them. As pressure builds in the middle of the cylinder, the flared edge of the seal tends to get more flared and make a stronger seal.

If you are working on an older foreign car that uses Lockheed or Bendix brakes, you can probably find kits to rebuild the cylinders at a regular parts store. The parts man has to be cooperative. I was able, for example, to get wheel cylinders for a '58 Volvo which were nominally listed for a Studebaker. The critical dimension is the diameter of the cylinder bore. The sizes increase in 1/32" increments.

The rear-wheel cylinders are generally smaller than the front cylinders. The brake shoes are generally smaller as well. The purpose of this difference is to balance the braking between the front and rear; as more of the weight is transferred to the front wheels, more braking efficiency is required in front.

If a wheel cylinder fails shortly after a brake job has been done and you don't want to replace the brake shoes, you can soak them in trichloroethylene for an hour or two. This will restore the brake shoes to a useful condition.

Brake Fittings and Flexible Hose

If the wheel cylinder has to be replaced, brake fittings has to be removed. Gas-line fittings have to cope with pressures up to 35 lbs. Brake-line fittings have to be removed. Gas-line fittings have thousands of pounds! The brake lines connecting the master cylinder with the wheel cylinders are steel. The ends are double flared by law. This flared end is pressed by the flare nut against a brass seat in the bottom of a threaded hole of the wheel cylinder or the master cylinder (I have purchased a replacement cylinder without this brass seat). Brass is used so it will conform to the steel flare. On older vehicles this seat may be damaged due to overtightening allowing a slow constant leak.

Loosening the flare nut, as it is called, should be done with a flare-nut wrench. A regular wrench will often round off the corners of the brass flare nut. If you are disconnecting the brake-line connection to the front wheels, notice that a flexible hose is part of the connection.

This is potentially the weak link in the brake system. If it is cut or cracked, replace it. This piece will burst if weakened and will do it at just the time it would be most dangerous—as the brake is heavily applied. When you remove this hose, remove first the end with two hexagonal fittings. Hold the larger fitting with a wrench while turning the smaller with another. The opposite end can now be unscrewed. There is another flexible base in the rear on a conventional rear-axle car.

When removing metal brake lines, bend them as little as possible so they will line up with the female fitting upon reinstallation. Start the flare nut in by hand to be sure it is starting properly. If it cross-threads, you will *have* to replace the wheel cylinder. It may be necessary to do some finagling, bending and moving around to see things at different angles and even some cursing to get it started by hand, but do it. And once it is in don't overtighten it. Rarely will additional tightening stop a leak in this type of fitting. Replacement steel-brake lines, already flared and fitted with flare nuts, can be purchased at the regular parts store for common American cars. K-D makes a tool that can do both single flares (used on gas lines) and double flares.

Backing Plate

Check the backing plate. If it is loose, tighten it. There are two or three raised rubbing surfaces on the backing plate. They lie along the arc where the metal shoe will be when it is replaced. These must be very lightly greased so that the shoes move on them with ease. A special white grease, called Lubriplate, can be used here and on the brake adjuster. If the rubbing plates are heavily grooved, the shoes can cock, causing the drums to be badly scored, not to mention a rather short brake-shoe life. You could grind out these grooves or get new backing plates.

New Brake Shoes and Drums

Brake shoes come either bonded (glued) or riveted. Some say riveted shoes are stronger, but even rivets can get loose. I have never had either type come loose. The bonded type is usually a little cheaper. The riveted shoe is generally the top of the line and usually has a better lining—more heat resistant. Therefore if the shoes or drums show evidence of burning use the riveted kind.

The shoes as they come in the box are actually unfinished. A finished shoe is one which is arced (ground) to fit your drum exactly. This gives the shoe and drum maximum contact area. If the shoes are put on without this being done, you may loose as much as 30% of your braking potential. A finished shoe is also chamfered along the short straight edges, which is supposed to forestall "brake chatter." Chatter has never happened to me when I failed to chamfer them, but it might solve the problem if it arises.

Shoes ordered from Whitney in Chicago run cooler because of a copper mesh imbedded in the asbestos. The shoe material can vary. If it is softer, it will brake quicker, but also wear out sooner. Sometimes if the brakes are running hot, a resin will boil out of the brake-shoe pads and glaze the surface of the shoe. If the brake drum is glazed or scored, it should be turned. "Turning" is a process of cutting down the braking surface of the drum until it is smooth and concentric. A drum slightly out of round will be remedied by this operation.

If this were all there was to it, it would be simple. However, besides providing a braking surface for the shoes, the drum must also dissipate heat. The more metal you remove from the drum, the smaller will be the heat load it can handle. Aluminum drums are the best for this purpose but are much more expensive to make. As brakes get hotter, a gas forms between the drum and shoe and soon the shoe is skating on the drum: No braking occurs. A state law in California limits turning of drums to a certain amount. You have a dilemma here: You don't want to remove any metal, but you want to have a smooth surface for the brake shoe to contact. If too much metal will have to be removed to get the smooth surface, your machinist will tell you. Usable drums can be purchased at the junk yard at 1/3 the cost of new ones; of course they may have to be turned.

This brings up a problem one always has to face with rebuilding. Within a reasonable cost you cannot expect to rebuild a system back to new. The best you can hope for is 90% and often you will have to accept less. Getting that last 5 or 10% may cost as much or more than the cost of the first 90%. If you have the money to spend, it is possible to improve even on factory standards, but I will not consider that here.

The new shoes can now be mounted. Use the hold-down springs to position them. If the linings are of different length, the shoe with the shorter lining goes on the front or leading position. If all four wheels are being relined the narrower shoes go in the rear. Make sure the actuator pins are straight. Hook on the return springs plus any necessary spaces or adjustment linkage. If you forget the arrangement, *Chilton's* can help you. I hook the return springs into their hole in the shoe, then with a medium-length screw driver pry them up and over their backing-plate post. If the spring is loose when mounted or quite stretched then you're in the wrong hole. A spring when mounted into its normal rest position, whether in the brakes, the gas pedal, or any other place, should be just barely extended. The spring generally should also have free straight movement. A spring shouldn't be rubbing anything, except for the rear-brake spring, which is purposely clipped onto the parking-brake line; and in some cases the bottom adjusting-screw spring, which may double as a device to hold the adjusting screw in place.

The expanding star-wheel brake-adjuster screw should have a little Lubriplate put on the threads and under the rotating cap, then be screwed into its shortest position. It is then installed at the bottom of the shoes with the star wheel opposite a hole in the backing plate. This is where adjustment will be made later. If you plan to remove the self-adjusting linkage, replace the bottom automatic adjuster spring with an ordinary adjusting spring. The parts store will have one, but you will have to figure out what size you want. I will defer the discussion of reinstallation of the drums, adjustment, and brake bleeding until I have discussed disc brakes.

Disc Brakes

Disc brakes are much simpler than drum brakes, hence fewer things can go wrong. The pads wear out. You can replace them by jacking up the car and removing the wheels. This exposes the caliper. Remove two pins held by spring clips and a tensioning spring. The pads can now be pulled or levered out. Many new pads have a groove cut in them. When the pad has worn down so much that the groove has disappeared, replace it. Otherwise replace the pads when only 1/16" or less remains.

Caliper Cylinders

With the pads removed, check for any fluid leakage. Calipers rarely need to be rebuilt, perhaps because the caliper pistons move so little. I have seen twenty-year, thousands-of-miles-old calipers that worked perfectly. To rebuild them, the calipers have to be removed. Disconnect the flexible hose then unbolt the calipers. They may be shimmed, so notice how many shims there are and where they are. The makers would rather you didn't separate the halves because there might be an alignment problem.

If you have compressed air you can blow a caliper piston out while blocking the other in place. Look for a seal at the bottom of the cylinder when the piston is removed. Clean out the disassembled piston with brake fluid and install new gaskets.

If the cylinder appears damaged, replace the entire caliper. Apply a thin coat of brake-cylinder lubricant to the parts, then install the piston. You may have to wedge it in using a large screw driver and a block of wood. With the one piston rebuilt, move to the opposite cylinder. Blow out the remaining piston and repeat the steps.

If no compressed air is available, you can separate the caliper into two halves. It may take a special wrench. Be careful to keep the mating surfaces of the two halves clean. With the halves separate, you can use a special caliper piston puller to finish the job. If you get brake "judder" upon reinstallation, just replace the caliper with one from the junk yard.

Whatever was mentioned for drum brakes about brake fittings and flexible hose applies for disc brakes. No backing plate is used in a disc-brake system, so no mention of it will be made.

New Pads and Rotors

New pads need no finishing. I spray the metal back of the pads with a Permatex product called *Disc Brake Quiet.* They can be installed directly once the caliper pistons have been retracted with a large screwdriver or brake-adjusting tool. These tools simply act as levers to push the piston in. Be careful not to damage the rubber dust seal. Brake fluid gets forced back through the lines into the reservoir on the master cylinder, which will overflow if it was full to begin with. If retraction of the pistons is difficult, try removing the cap on the reservoir.

If the old pads were worn to the metal, the scores were probably made in the rotor. It should be removed before the new pads are installed; but first remove the caliper. The rotor has a minimum thickness below which it shouldn't be turned. The machinist will decide for you. Junkyard rotors could be used, if you can find a rotor metal still on it. Thin rotors will result in brake fade.

A pulsing felt through the brake pedal can sometimes be corrected by turning the rotors. To identify the offending rotor, jack up the front of the car and spin each wheel. The wheel which gives a slight scraping in a rhythmic pattern is the one to remove and have machined.

Reinstallation of Drums or Rotors

New or cleaned axle bearings in the front are repacked with wheel-bearing grease. Disc brakes need a special type of high temperature grease. The drop point of a grease refers to the temperature at which it melts and runs out of the bearing. I use disc-brake wheel-bearing grease on all applications. It's the best product.

Repacking the bearings means taking a cherry-sized lump of grease and working it into the rollers. The grease you can work in is essentially all the lubrication these rollers will get until they are repacked again. Place the inner bearing into its race and work another dab of grease into the exposed roller ends. Carefully install the inner seal. It can be pounded in with a hammer and a large socket that has a diameter slightly smaller than the seal.

Now pick up the drum or rotor and place it on the spindle. Follow the previous instructions for working grease into the outer bearings. Install it on the spindle, then add grease, the washer, and the axle nut. The function of this nut is unique. It doesn't simply hold something together; it does so at just the right pressure. Turn this nut with a large crescent wrench while spinning the drum or rotor. As the nut gets tighter, notice if the drum or rotor slows down, or gets harder to turn. Snug up the nut, then back off to a point where the resistance just begins. If no resistance to spinning was ever noticed, back it off a quarter of a turn anyway. Just the slightest resistance in the setting puts a preload on the bearing. This is desirable wherever a pair of tapered roller bearings are used on a shaft. If the setting is too tight, the bearings will grind themselves to death for lack of room for lubricant. A cotter pin is often used on American cars to keep the axle nut from turning. Foreign cars use lock nuts or nuts with a collar that gets staked.

When the wheel is put on again, rock it from top and bottom to see if there is any play. Turn the wheel to see if it spins freely and quietly.

Brake Adjustment

No adjustment is required for disc brakes. Drum brakes are adjusted with the wheel off the ground. Somewhere in the back of the backing plate, near the bottom, is an oval hole for inserting a screw driver or a brake-adjusting tool that will be used to turn the star-adjusting screw. On VWs and a few General Motors cars, a hold in the drum provides access to the star adjuster. Sometimes a metal plug has to be knocked out of the drum.

The adjuster was reduced all the way to begin with, therefore, you won't go far in the wrong direction. Continue to expand the adjuster screw until the wheel, which you are spinning, locks up. Then back off the adjustment until you hear only a slight scrape usually after two to four partial turns. With automatic adjustors a long, thin screw driver must be inserted through the backing plate to push the adjusting lever aside. This lever impedes the movement of slackening the brake adjustment. When adjusting the remaining brakes, try to slacken off the adjustment to the same degree. I remember a '50 Olds that had to be backed off fourteen partial turns or the brakes would lock. If the car doesn't roll too well, the brakes may be too tight. Tight brakes will burn.

Brake Bleeding

New brakes may have to be readjusted after a thousand miles or so if the shoes weren't arced or the drums turned. Then one or two adjustments are in order before the next brake job unless they are self-adjusting.

If the hydraulic system is interrupted in any way, brake bleeding will be necessary. You force fluid through the lines by the brake pedal activating the master cylinder. This will push air in the line ahead of it and out of the bleeding ports in each wheel cylinder, one at a time. Bleed the cylinder furthest away from the master cylinder first, the right rear, moving progressively closer. If only a pair of wheels are being repaired, only bleed those brakes. If the pedal is still spongy, then do then all. We want in the lines only something that is incompressible and doesn't expand much with heat, that is, something inert. If too much air is in the lines, you will get air compression when you push the brake pedal, not brake stoppage. K-D makes a nice one-man brake-bleeding tool. Find the right-size box wrench to fit the bleeding-nipple hex or a socket from the 1/4"-drive socket set and open the nipple half a turn. Place the hose of the bleeding tool over the nipple leaving the box wrench in place if possible. Go up front, check to see if the fluid reservoir is full. Remember to use Girling or Castrol brake fluid on Girling brake systems. Depress the brake pedal watching the bleeding hose if possible. You can use a two-by-four to depress the brake pedal from outside the car. If you can't depress the pedal, open the bleeding nipple a little more. There should be spitting and hissing as air is forced out. When you get two or three clear streams of fluid, stop pumping the brake pedal. The one-way valve at the end of the bleeding tool keeps air from being sucked back into the line. Go back to the bleeding nipple and close it. I do not reuse brake fluid, so it goes on the ground. Brake fluid is very corrosive to paint, by the way. Having done all the brakes, I try the brake pedal. It should be high, because the brakes were just adjusted and they should not be spongy. If they are, try bleeding again. I have run into brake systems that had to be done under pressure. This is very rare, however. If the pedal is first hard and high but slowly drops to the floor, either one of the nipples is still slightly open or the master-cylinder seals are shot, and the master-cylinder should be rebuilt.

Master Cylinder

If there is a leak from the plunger end of the master cylinder or the plunger is just pushing through the brake fluid, the brake pedal going slowly to the floor, it's time to rebuild the master cylinder.

Sometimes a pressure switch for the stop lights will leak causing a short and definitely a loss of brake fluid. This can be remedied by replacing the switch. Have the new switch in hand so it can be screwed in directly before all the brake fluid runs out.

To remove the master cylinder for rebuilding, first loosen the flare nuts with a flare-nut wrench. Remove the lines, trying not to bend them or to allow fluid to run out. Disconnect the wires running to the pressure switch, if there is one. Then remove the mounting bolts. If the fluid reservoir is not attached to the cylinder itself, all the fluid will run out as you disconnect the pipe running between the two.

When the clip at the end of the cylinder is

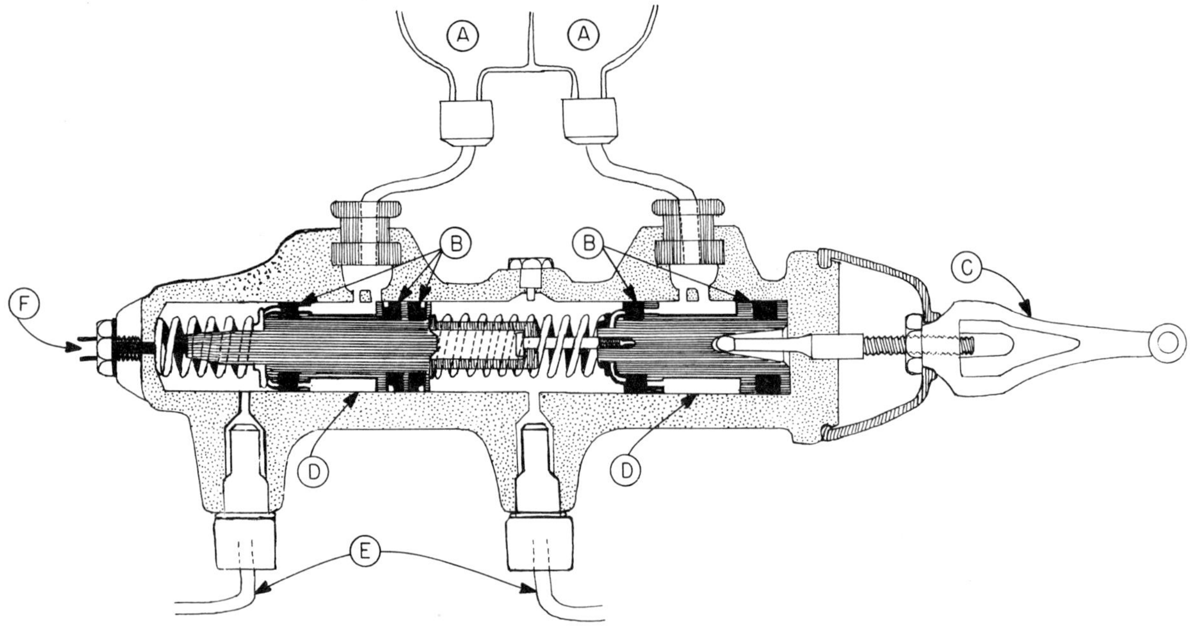

A Brake Fluid Reservoir — C Brake Pedal Linkage — E Outlet to Wheel Cylinders

B Rubber Seals — D Pistons — F Stop Light Pressure Switch

removed, the piston may pop out. If it doesn't, and if it's a dual braking system with two reservoirs then there is a stop screw that should be removed somewhere along the body of the piston. The second piston of a dual system may have to be knocked out with sharp blows of the cylinder against the cement. Take a close look at these pistons and remember which goes in first. Examine the cylinder bore to see if there is any scoring or pitting. If there is any scoring or pitting, replace the whole unit.

My only failure with a kit was a thirty-year old vehicle. I clean everything with trichloroethylene, then steel wool the bore, and finally clean everything with brake fluid. I don't hone the cylinder but some mechanics insist on it.

When the new seals are reinstalled, they can be lubricated with brake fluid or brake-cylinder lubricant, which may come in the rebuilding kit. I try to start the flare nuts by finger before I mount the master cylinder, so that alignment of the brake lines and the threaded fittings is easier. Then I mount the master cylinder.

When you have everything reconnected, tighten down the flare nuts, the back them off two or three partial turns. Have a friend pump the brake pedal. When all the hissing and spitting around the fitting ceases and nothing but fluid is leaking, tighten the flare nut again. If this doesn't work and the brake pedal is spongy, bleed the brakes from the wheel cylinders. Refill the brake-fluid reservoir.

In a tandem-brake system, if the brake warning light is on, temporarily open up the flare nut on the brake line of the master cylinder going to either the front pair of brakes or the rear; whichever pair *didnt* fail. Press on the brake until the warning light goes out. You are recentering the switch which turns it off. Then tighten down the fitting again.

Road Test

One should road test just about any repair made on a car, but for brakes this is most important. Get in the car and press on the brakes as hard as possible to see if anything bursts. Pump them to see if air is in the system. Hold the pedal down under heavy pressure for a while to see if it goes slowly to the floor indicating a leak.

Then take her for a drive. Pump the brakes while moving to see if strange things are happening. On a safe road get up to thirty or forty and do a panic stop; enough to see whether the brakes pull to one side or another. If you have trouble rolling or smell burning, probably the brakes are adjusted too tight.

Chapter 14:

The Clutch: Driveshaft and U-joints

We are all working together to one end, some with knowledge and design, and others without knowing what they do.

—Marcus Aurelius (A.D. 121-180)

The clutch job can weaken your resolve to be a mechanic. Talk about dirt in the eye, the dangers of working under a jacked-up car, and bolts that are nearly impossible to reach; this is it! I generally charge a higher hourly rate to do clutch jobs. One has to be properly motivated.

The function of the clutch is to slip. But to slip only on engagement of the engine to the rear wheels through the drive train. This slipping is necessary because the engine is not able to carry a load below 1,000 or 1,200 r.p.m. If the engine were instantaneously and directly linked to the stationary, rear wheels the car would lurch forward; bad for the neck and bad for the machine. Like the brakes, then, which are another slip-friction device, the clutch friction disk wears out. Before we go further let's learn to "read" the clutch pedal.

Reading the Clutch Pedal

The 1/4" of movement in the pedal should be easy because it merely reflects the movement via the pedal linkage of the throwout bearing from its rest position to engagement with the clutch pressure plate. You are here working against the clutch-pedal spring. A change in amount of movement is the only external adjustment that can be made on the clutch without taking it apart. The adjustment for this is generally in the length of the rod that engages the throwout fork just outside the bellhousing of the transmission. Only in one car, the English Sprite, have I encountered a nonadjustable linkage; the dealer does supply different-length rods, however.

The absence of a space, in the beginning of the pedal throw, may be due to a tight adjustment. This tightness may be accompanied by a whirring sound of the throwout bearing spinning with the clutch and by a slipping clutch. Less frequently the absence of space may be due to a broken clutch-pedal return spring.

Too much space might make the movement of the throwout bearing too short to disengage fully the clutch from the flywheel, making it either impossible to shift unless the engine is off, or causing a grinding of the gears.

If hydraulic linkage is used for the clutch rather than mechanical (as in many foreign cars), then an additional small movement is present in the clutch pedal. When the pedal is replaced, a small space should be present between the pedal linkage and the piston of the clutch master cylinder. This is similar to the situation with the brake master cylinder. This space is generally not adjustable nor does it need to be.

To determine the place where the clutch engages, turn on the engine, then with the pedal pushed to the floor, engage first gear. Slowly let the pedal out and notice when the car begins to move that is, when the clutch begins to engage the flywheel. I'm afraid this doesn't tell you too much. A worn-out clutch might only engage when the pedal is nearly fully released, but then it might not. In a diaphragm-type clutch (the type used in most modern cars), I perfer to see engagement within the first half of the upward pedal movement. If the throwout bearing has too great a throw, it can push the diaphragm too far, springing it, and destroying the pressure plate.

If the clutch pedal screams when you push it in, the throwout bearing is shot. If the pedal goes to the floor easily, the clutch linkage is either broken or leaking (hydraulic) or the cross shaft inside the bellhousing is broken.

Clutches usually need to be replaced anywhere from 40,000 to 100,000 miles. The symptoms of a worn-out friction disk are the engine racing ahead with a loss of power on a hill or at freeway speeds. High-performance engines will produce the symptoms on a fast getaway. Eventually the car will not be able to make it up the slightest grade.

If the pedal pulses during engagement, look for a damaged pressure plate or possibly a bend in the input shaft of the transmission. Clutch chatter may be caused either by oil leaked on the friction disk or by impeded movement of the clutch linkage, which hinders smooth clutch engagement. To solve any of these problems (except the problem of a maladjusted clutch linkage) the clutch must be disassembled.

Clutch Linkage

Two types of linkage are used on clutches. Mechanical linkage is used on most American cars, hydraulic linkage on most foreign cars. Mechanical linkage is quite straightforward; the only problems I have encountered with it are a loose shaft on which the pedal is mounted. It will be hard to get parts for this on older cars and it is only on older cars on which the problem occurs.

Hydraulic clutches, other than hydraulic brakes, have no flowback valve in the master cylinder. Also the clutch system isn't exposed to nearly as much pressure as the brake system, hence plastic lines are sometimes used. The simplest problem is leakage at the slave cylinder, accompanied by loss of fluid in the reservoir. A leak anywhere in the system can be quickly identified by having someone depress the clutch pedal while you watch the linkage at the bellhousing. If the throwout fork doesn't move at all, the fluid has probably completely leaked out of the hydraulic system. If the throwout fork moves, but slowly retracts even though the clutch pedal is still being depressed, there is a leak in the hydraulic system that is still being made up by replacement for the fluid reservoir. If no fluid is visibly leaking from the slave cylinder attached to the bellhousing, then the seals in the clutch master cylinder are leaking. This last leakage may manifest itself as gear clashing during shifting, which can be avoided by pumping the clutch pedal first, then shifting. Advice on rebuilding hydraulic systems may be found in the *Brake* chapter.

Besides linkage differences, there is a difference in the plate that presses the friction disk against the flywheel. Most newer cars use a diaphragm type. It is easier and quicker to make, but in my opinion less satisfactory. I will have more to say about it when we have the clutch disassembled.

Special tools are required to release the pressure on the pressure plate, however. Consult a shop-manual abridgment to see what type of tool is needed. I fabricated one out of hardwood for a

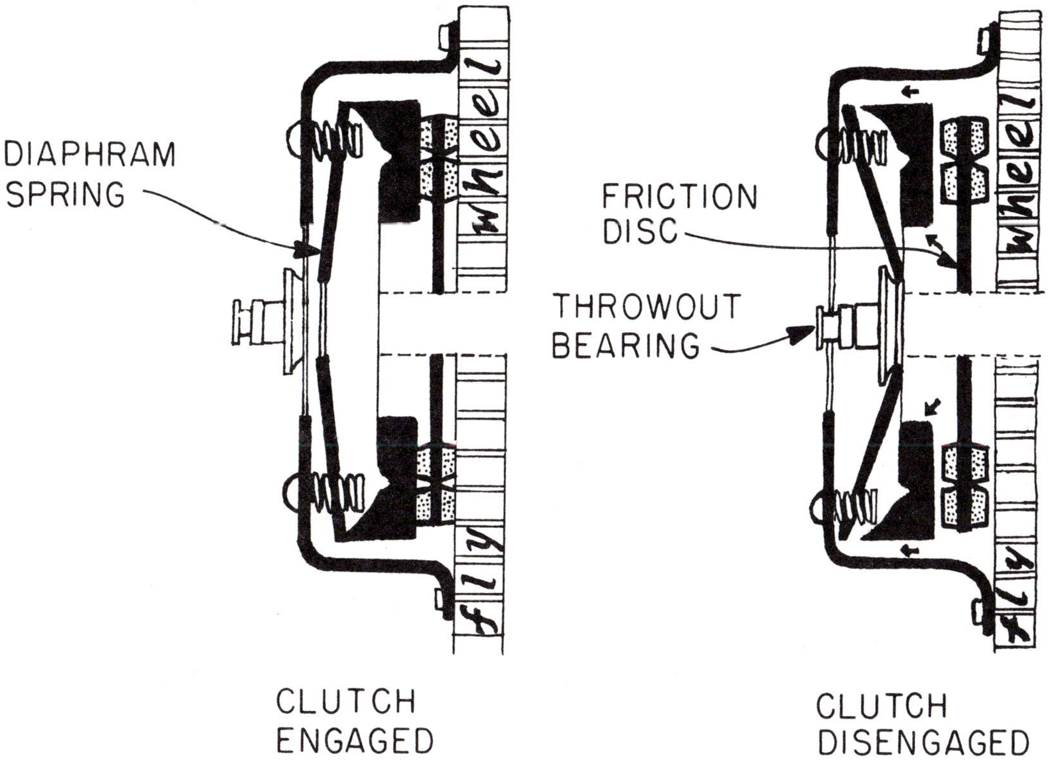

Clutch Disassembly

Most clutches are removed by taking out the transmission and bellhousing together from underneath. The bellhousing encloses the clutch. On some trucks the bellhousing is left in place and the clutch removed from the bottom after the transmission is out. The bottom bellhousing bolts on the Chevy half-ton truck are inside the bellhousing. The bottom bolts on some Internationals are removed from outside the housing in a spot that is hard to believe. All I can say is, good luck! On many smaller cars (English Sprite, Datsun 1600 sports car) the engine and transmission come out as a unit. For these cars consult the *Engine* chapter. All front-wheel-drive cars on which I have worked have allowed for the clutch to be removed while the engine and transmission are in place. Saab 99, but I was not able to find the necessary special screws needed for the Austin America. The dealer probably won't sell you the special tool.

I like to get the car off the ground 10 inches or so, and on four jack stands, when transmission removal is the preferred method to reach the clutch. This is enough room to have freedom of movement while lying on a creeper. One generally disconnects a speedometer cable, wires for a backup-light switch, and clutch connections. The shift linkage to the transmission has to be disconnected. On floor-shift cars the shift lever inside the car has to be removed. There may be bolts holding this lever in place. On some Japanese cars the collar around the shift lever is a spring-loaded bayonet fitting (as on many light bulbs), which has to be depressed and turned to be removed. This type is hell to get back together. Some cars require disconnecting the exhaust pipe from the exhaust manifold. Plan to replace the flange gasket between the two.

The driveshaft, which is discussed in detail at the end of the chapter, has to be removed. When the shaft is pulled from the end of the transmission on some cars, transmission fluid will leak out. This can make things very messy. I recommend draining the fluid by about half before you get to this stage. In fact, draining all of it and putting in fresh 90-weight oil would be a good idea.

With a transmission jack in place to support the transmission, the bellhousing bolts and any transmission supports can be removed. Two of the lower bellhousing bolts also hold on the starter motor. The starter motor can often be left approximately in place. The top bellhousing bolts are sometimes easier to reach from the engine compartment, but often the best approach to these very inaccessable bolts is to allow the engine and transmission to drop down a few inches after the transmission supports are freed. The transmission then has to be pulled back a few inches to allow the input shaft to come free of the clutch. Don't let the transmission hang by this shaft.

Sometimes it is easier to unbolt the transmission from the bellhousing and remove this unit first. Then the bellhousing will come off. The Volvo 122 series is done this way. *Chilton's* gives a step-by-step procedure for any car, but don't assume it is the best method for you.

When the bellhousing is removed, the pressure plate is exposed. It is mounted on the flywheel by six bolts. Loosen each of these a little at a time, around and around. This method relieves the pressure being exerted by the pressure plate evenly, which helps to avoid warping. The flywheel will tend to turn when these screws are being loosened. This tendency can be overcome if you place the ratchet handle across the center axis of the flywheel and pull, or strike the handle of the ratchet smartly with the palm of the hand. Before the pressure plate is completely removed it should be marked, so that if it is reused it will go back in the same position.

Balance

Anything in the drive line which goes around, from the crank pulley to the wheels, should be balanced for best results. Balancing reduces vibration. Vibration kills machinery. Therefore when the flywheel is removed from the crank, the pressure plate from the flywheel, or the driveshaft from its couplings, they should be marked so they go back the same way.

It is alright if the parts go together 180 degrees out of phase, because the balancing process is bilaterally symmetrical. Many pressure plates or U-joints connections are designed in a way that only allows you to put things together in the same relationship or a half turn (180 degrees) away from the original relationship. If six bolts are used on a pressure plate, then six different positions are possible when you reinstall it. Two of these positions are acceptable; four are not.

If you are putting a new pressure plate on the flywheel, you won't know how it is balanced, so you have to take your chances. Look for drillings along the edge of the new plate. If you put them opposite similar drillings on the flywheel, you've done the best that can be reasonably expected.

Checking the Clutch

Take a look at the friction disk, which falls out when the pressure plate is removed. If it is oil soaked, check the seal where the input shaft goes into the transmission, and look behind the flywheel. There is a seal around the rear of the crankshaft. It may be bad if the main bearing clearance is too great. The crankshaft will pound out a new seal in this case, and the only good solution is to rebuild the engine. The oil leak is just as likely to be from a bad valve-cover gasket up above, so inspect that area as a possible solution to the problem.

Check the friction disk to see if the friction material is worn down to the rivets. If it is, the flywheel will be scored and should be resurfaced. Whenever you have access to the clutch, it is worthwhile to replace the friction disk whether it is completely worn out or not. You should teach yourself to recognize whether the disk is worn out, however, to be able to confirm or abandon your hypothesis about a bad clutch. Take a look at the new disk when you buy it, to compare with the old. A bad friction disk may be due also to loose rivets or a loose hub.

Check the pressure plate. Look for broken springs, cracks, and unevenness. Unfortunately it is nearly impossible to check this unit thoroughly without special equipment. Stand on the center ring to operate the unit. It probably won't tell you much but it might reveal a gross deficiency. On cheap cars I would replace a diaphragm pressure plate automatically. If the clutch problem was one of long duration or if there is evidence of burning by the smell and blue marks on the flywheel, replace the pressure plate. In high-performance cars, where great stresses are put on the clutch, replace the pressure plate.

I don't replace the pressure plate if it is a spring plate, if there is no evidence of damage, and if the engine is conservative.

Should you buy a rebuilt pressure plate or a new one? One out of fifteen or so rebuilt pressure plates fail in my experience. Since this ratio can vary widely from area to area, one takes a chance with a rebuilt unit. New units, of course, cost twice as much or more. If you are working for others give them the options ahead of time.

Spin the throwout bearing. If it is rough or noisy, replace it. Check the pilot bearing in the center of the flywheel. The nose of the transmission input shaft goes in this. On some cars it's a brass bushing, on others a roller bearing. A high-pitched moan when you start up in the morning or when you shift may be a result of a bad bearing.

Check the flywheel. Look for scoring and bluing. Scoring is mechanical abrasion, which grooves the flywheel surface; bluing is caused by overheating when the clutch slips. Scoring reduces the amount of surface area available for clutch engagement, and clutch life will be shorter. A fingernail-thin groove, even if it is deep, is not generally a problem; it is rather a general unevenness that calls for the flywheel to be removed from resurfacing. On the question of burn marks, my machinist claims that they can cause clutch "judder" (shake). On the other hand, I did a clutch on a Reo school bus which had the most blued flywheel I've ever seen. In addition, there was a network of fine cracks across the surface of the flywheel, not very deep. I didn't resurface the flywheel and the clutch operated properly for at least 10,000 miles before I lost track of the vehicle. The nicest jobs, however, are when the flywheel is resurfaced and everything else is new.

While we're here, the flywheel ring gear may be checked. The Lucas starters used on Triumph and Aston among others tends to chew up the teeth of the ring gear. Since the engine tends to stop just before full compression, only two or three spots on the ring gear will be destroyed, and the ring gear can be removed and rotated to get healthy teeth into the area of higher wear. If this is not possible, a new ring gear is needed. To replace the ring gear or resurface the flywheel, you have to remove the flywheel.

Flywheel Removal

To remove the flywheel with the engine in the car, it has to be chocked or wedged to keep it from turning. Sometimes a large screwdriver wedged into the ring gear and braced against the frame will do the trick. If the oil pan is off for some reason, the handle of the screwdriver can be wedged between the crankshaft and the crankcase. As a last resort, put the appropriate socket on a braker bar. Hold it in place on the nut with one hand and strike the handle with a sharp blow from a 2-lb. hammer. The momentary impact is usually sufficient to brake the nut loose.

VW and Porsche engines have to come out, if you want to reach the clutch and the flywheel. The flywheel is held in place by a large gland nut. It is called a gland nut because of the cavity or hole in the center for the nose of the transmission input shaft. This nut is torqued to over 200 lbs. so you have to have about four feet of leverage to loosen it. Flywheel chocks for 12-V

6-V flywheels are available at specialty and foreign-parts stores.

Flywheel bolts are special, so keep them apart. When you reinstall the flywheel, use locktite on the threads.

Putting it Together

With the transmission outside of the car, try the new friction disk on the input shaft on the transmission, to see if it fits and slides easily. Spray the shaft with McKays Graphite spray or put a *small* amount of high-temperature molybdenum grease on the shaft. Try the throwout bearing for movement. Some throwout bearings have to be driven off a sliding collar before the new one is installed. Try the sliding collar in place after the new bearing has been put on. Sometimes a little polishing with 320 emery paper is necessary to achieve smooth sliding.

The friction disk and pressure plate have to be held together as a unit with one hand, while the pressure plate screws are started by hand. I use Locktite on these screws. Be sure the hub of the friction disk (the raised middle) is pointing away from the flywheel. The springs in the hub are to help cushion clutch engagement. Some friction plates give quite a springy feel to the clutch.

Tighten the pressure-plate screws by hand until the friction disk can be moved by hand, just barely. The friction disk has to be centered with the pilot hole in the flywheel. Alignment tools are available for most clutch types at the parts store; some of them are excellent, some are worse than nothing. I usually do the centering by hand and eye. By eye I get the disk approximately centered; then with finger tips I feel around the edge of the friction disk where it abuts the pressure plate. There are three exposed areas around the circumference of the clutch. My fingers tell me if the disk is flush with the plate or at least evenly recessed all around. If you are working on a recessed flywheel, you have to get a pilot shaft to get the friction disk centered. Finish tightening the pressure-plate bolts in stages, round-and round. If the bellhousing is mounted separately, bolt it in place.

Now comes the moment of truth for this job. The input shaft of the transmission has to be eased in through the splined hub of the friction disk, and the nose of the shaft has to enter the pilot hole. This last inch is often hard to move. Move your jacks so that the bellhousing and its mating surface are parallel. This means the shaft should be nearly on the correct axis to enter the pilot hole. Get right into it—wrestle the transmission around pushing forward all the while.

If the bellhousing was mounted first, you have to aim the shaft to go through the throwout bearing as well. The advantage of this situation is that if you encounter a difficulty in moving the last inch, you can have somebody depress the clutch as you try to shove in the transmission. Depressing the clutch allows the friction disk to move. This little operation has brought me to tears of rage and despair more than once.

With the transmission popped in, it is merely a matter of reversing the steps taken during removal. When reinstalling exhaust-system parts, get all the connection points loosely reconnected and therefore aligned, then tighten everything down.

Clutch Adjustment

The clutch has to be checked for adjustment. Grab the throwout fork just outside of the bellhousing and move it back and forth. You can look up the manufacturer's suggestion for the amount of play here, but ¼" will do quite well. Adjustment is made by changing the length of the rod that engages the throwout arm. A lock nut is used here. Make sure it is tight when you get the adjustment you want.

When you have put everything together but get nothing but a whirr after you engage the clutch—jack up the car and start over. The car company probably changed the clutch in the middle of the model year and you didn't check the new friction disk with the old, and didn't double check by trying the new one on the shaft with the transmission out. If it lamely works but makes strange sounds, the friction disk is in backward.

Driveshaft

In conventional cars the mainshaft of the transmission is connected to the driveshaft, which in turn is connected to the differential. Driveshafts have at least one universal joint and may have as many as three, they may be exposed or run inside of a tube.

The symptom of a bad universal joint is play in the drive line—especially when you shift between first and reverse, and get a jerk. You may also get a loud thumping on acceleration from a standstill. Jacking the car into the air, you can roll under and grab the driveshaft and twist it back and forth looking at each universal joint as you do this. There should be no movement at all in the joint.

Many older cars have a closed driveshaft. The front of the shaft is splined and fits into a cup. The universal joint is just before this connection. The standard procedure for braking the driveshaft connection to the transmission is to jack up the rear, disconnect the hydraulic and handbrake connections, and drop the rear end away from the suspension. The housing joining the front of the axle to the transmission is disconnected, and the rear end is pulled straight back. In old Chevys, I have been told but haven't had the opportunity to try it myself, that, if you pick up the rear end high enough, the driveshaft will pull out of its connection without the need for disconnecting and removing the rear end.

Once the closed driveshaft is disconnected, the universal joint is available for removal. In this design the joint is lubricated by transmission fluid. Be sure to renew all the oil seals because fluid can leak back through the driveshaft tube overfilling the rear end, which will leak out of the rear main seals into the rear brakes.

Of the exposed-driveshaft type, the three universal-joint Volvo driveshaft gave the most trouble in my experience. I don't have a Volvo customer who hasn't replaced at least one of his or her universal joints (spiders). The local dealer always replaces all three of the joints if any one of them goes bad. I still only replace those that are bad, but I always tell the customer that it's not uncommon to have one or both of the others go bad shortly.

The driveshaft is supposed to be balanced, which means it should be put together the way it was taken apart, or 180 degrees turned from the original position. Many driveshafts can only be reconnected in the original position or 180 degrees from it. A small vibration introduced into the driveshaft can hasten the destruction of universal joints, and even possibly the transmission or the differential gears. In fact, any part of the driveshaft may telegraph destructive vibration to any other part. Therefore if a car shows a tendency to go through U-joints, consider having the driveshaft balanced. If you experience a symptom like a bad U-joint but can't see it, it may be a bad bearing on the pinion shaft of the differential.

Universal Joints

Universal joints may have either two press-fit bearing cups and two U-clamp-fitted bearing cups, or four press-fit bearing cups. The U-clamp-fitted types require only the removal of the U-clamps to separate the joint and free two of the bearing cups. The other two are pressed out in the manner described below for the four press-fit type which is coupled by nuts and bolts. These nuts and bolts must be used with lock washers or, better, Locktite.

First remove all snap rings from behind all bearing cups. Then, with the yoke of one shaft of the joint laid across the supported on top of the open jaws of the vise, drive the driveshaft down with a hammer to force one of the bearing cups out of its holder. Turn everything half-way around and do the same to the opposite side. The cups may be completely removed with a pipe wrench, or maybe a small punch, from the inside. This job would be easier if you could heat up the whole joint to 90-100 degrees. With at least one of the cups completely out and the other most of the way, you should be able to separate the joint.

The remaining two cups could be driven out of their holders by pounding on the exposed arm of the universal joint. Never reuse a universal joint after it is removed. Thoroughly clean the two yokes in solvent. Examine the joint to verify that it was damaged. Good bearings are clean, bright, and unpitted.

Take the new joint and remove two opposing bearing cups. Check whether they are completely grease filled; if not put a fingerful of chassis grease in each cup. Heat up the two yokes if possible, to make the fitting easier. Put the joint without the two cups into one of the yokes. Take the two cups and place them in the appropriate position to squeeze into their holders with a vise. Do this slowly being sure each side of the joint is easing into its bearing cup. If the bearing cups get cocked in their holders and aren't sliding in, shift the position in the vise and squeeze the high side. These cups can be driven into place with a hammer and a drift punch or a socket of the appropriate size, but you have to be careful that the joint is held in each cup as it is being driven in. Otherwise the roller bearings will come out and you'll have to start again. When the cup is flush with the holder, a socket can be used to force the cup further along to the point where the snap ring will fit. Follow the same procedure with the other two cups and yoke. If the other two cups are held in place by U-clamps, you might put tape around the free bearing cups to keep them in place until the driveshaft is in position.

Sometimes you can't drive the cups in far enough to get both snap rings in at the same time. This means that the yoke has been pressed together slightly. You can start from scratch with a new joint, correcting the yoke somehow. My recommendation is to center the joint as well as you can, then run some red Locktite around the edge of the bearing cup to hold it in place. If the fit was tight to begin with, that will probably hold it. You might check the joint after 500 to 1,000 miles to see if the bearing cup is moving. This is one of those occasions where you have to make a thing work whether it wants to or not.

Split driveshafts have a hanger device in the middle. It may include rubber grommets or a ball bearing. Check these for wear because they may have been the cause of a universal joint going bad. Bad U-joints often cause the destruction of the rear seal of the transmission or the seal in the nose of the differential. Any sign of leaking at these places can be corrected by replacing the seals while the driveshaft is out.

Chapter 15:

Engine Overhaul (Bottom End)

Fascinated by the delicate precision and sheer power of the devices he has invented, man is disposed to use them for doing whatever by their aid can be done, in the confident expectation that what can be done with such clean efficiency must be worth doing.

—Carl L. Becker (1873-1945)

Engine overhauling produces the greatest satisfaction for the mechanic and is the best testimonial to his work. The satisfaction comes from taking apart a machine consisting of hundreds of parts, examining them all with a practiced eye and having the appropriate parts renewed by various specialists. The parts are then reassembled, with very tangible results to feed the mechanic's anxious curiosity. The parts man or the machinist don't receive this clear positive or negative feedback.

You will probably get the best wages for this work, or save the most money doing the job yourself. The engine is generally the most rebuildable part of the car. The parts to do it are usually the last to become unavailable (body parts are the first). Engine rebuilding is often the most economical solution if compared with the cost of replacement of the whole car. Let's first see when engine rebuilding can be expected.

When to Rebuild

There is no magic about mileage—an engine lasts only so long. A small short-stroke engine that has to revolve more to go the same distance as a long-stroke engine, will obviously not last as long. A small cheap engine that may need a valve job at 50,000 to 90,000 miles will probably need total overhaul from 100,000 to 150,000 miles (the average driver puts 12,000 miles on a car in a year). A better car will last from 100,000 to150,000 miles before a valve job and up to 250,000 miles before a total overhaul is necessary. Engines are designed to be used in production for 5 to 10 years, with an increment of performance squeezed out by boring and stroking every few years. Boring increases cylinder size but creates heating problems. Stroking makes for higher compression but increases stresses on connecting rods and bearings. Therefore an engine at the end of its design period will have better performance and a shorter life expectancy. Other factors to consider can considerably shorten these outside limits: Engines mismatched for power versus body weight or with an inappropriate rear-end ratio for the type of vehicle will not last as long. The obvious example is the VW bus which has an average life expectancy of 50,000 to 70,000 miles.

People who don't change their oil regularly will shorten the life of their engine, because sulphuric acid builds up, which shortens ring and bearing life.

An engine with an overheating problem won't last as long as a healthy engine.

Finally, an engine made on Wednesday will probably be better from the outset than one made on Friday afternoon to the extent that humans are used to assemble them.

So much for the engine as it comes off the assembly line. Rebuilt engines are another story. There are as many ways to rebuild an engine as there are ways to skin a cat. Short of blueprinting an engine, which is done in racing, the best you can expect from a rebuild is 70-80% of the orginial factory life, with a lot of ways to go wrong. With every subsequent rebuild, subtract another 10-20% and expect a greater cost for parts. If the first rebuild held up well, junk the car as the whole, when the time comes for a second engine rebuild; the body and suspension are getting tired, among other things, and the car would probably be in need of a lot of work. Junk-yard engines may be a good buy if you could be sure of what you are getting. Dealers steam clean the outside, so all engines look good. My perference is to rebuild an engine, but then I'm a mechanic.

The longest time I have ever heard of internal combustion engines running without major overhauls was 25 years. The engines were a pair of Caterpillar diesels (used to drive electrical generators), which were run alternately and continously, first at Yellowstone Park, then at Lassen Park. Since they were not turned on and off all the time, they experienced very few warmup periods. During the warmup period most of the wear of a car's engine occurs.

Signs of Trouble

If the engine seized on the freeway and it is impossible to turn the engine over by any means, then the signs are clear. If the oil-pressure light comes on at idle, and oil consumption has been high, the main and rod bearings are probably worn out. If on opening the hood you discover the engine compartment covered in oil, a ring is probably broken and compression building up in the crankcase is blowing oil out through all the seams (but check for a blocked PCV). Hollow knocking sounds could come from the crankshaft, and sharper rapping noises from the connecting rods. If the rear main seal is badly shot (oil leaking profusely in the area of the clutch) the main bearings are once again implicated.

Performance from the engine itself mostly depends on the condition of the rings, head gasket, and valves. You can read 180 lbs. in all cylinders and yet throw a rod tomorrow. For a consideration of poor compression consult the *Valves* chapter. If a compression test reveals that the rings are leaking, if oil consumption is high, if unseemly noises are coming from the crankcase, if oil pressure is bad, and, most important, if the mileage on the engine is high, consider rebuilding the engine.

One comforting thought is that rarely will you open an engine and find no damage. It happened to me only once, and on my own car. The accelerator-pump nozzle fell out of the carburetor into the number-three cylinder. The resulting poor performance and sound convinced me that I had lost the exhaust valve. Imagine my surprise to discover a completely clean standard engine. It's hard to make an accurate diagnosis of sounds, by the way, if the exhaust system is noisy.

One of my favorite tricks to test quickly for over-all engine condition is slowly to turn down the idle screw. See how far down you can go without the engine stalling. Dropping the idle may reveal some interesting noises. Great differences in compression will show up as sharp roughness. Conversely, a common device to cover up engine deficiency is to set the idle speed high. A crude method of measuring main-bearing clearance is to run the engine till hot, turn off the key, then immediately turn on the key again without turning on the engine. See how long it takes for the oil-pressure light to come on. If it comes on right away the oil is leaking out of the bearings rapidly, and wear is probably great. If it takes five to ten seconds for the oil pressure to drop and the light to come on, then the bearings are not extensively worn. The time will also depend on the weight of the oil in the engine and how hot it is. Try out this test on known engines many times before you rely on it for a judgment.

Another indicator of engine problems is to take the car for a run on the freeway to see if it overheats or smokes. If there is any doubt about the diagnosis, as there may well be, do a tune-up to look for further signs and to eliminate some of the more common sources of poor performance.

Very infrequently a small section of a main bearing will seize the main journal and will peel back. This usually occurs at freeway speeds with a loud bang and the engine stopping momentarily. If the engine frees itself and starts, again, pull off to the side of the road and let the engine cool before countinuing. I have over-hauled an engine 10,000 miles after this happened. The overhaul was occasioned by broken rings rather than a peeled bearing; therefore I assume the main potential problem is a bearing that doesn't free itself and an engine that remains frozen.

Generally it's better to do an engine overhaul just before something seizes or brakes rather than after, if it can be planned that way. It will cost more in the latter case. If you put a rod through the crankcase of a VW, you add a lot to the parts bill. One would like to rebuild an engine ideally at about 80% of its maximum expected lifetime. It is during the last 20%, when bearing clearance becomes excessive, that the crank will be pounding away creating most of the damage and a lot of the expense of an overhaul. In fact, if I get an engine that has been pounding for a long time (the oil light regularly coming on at idle) or if a connecting rod has been thrown through the crankcase, I try to get another lower-end assembly to rebuild.

Special Tools

Some special tools are necessary for engine rebuilding: A torque wrench; a large socket or two for the nuts that hold the fan pulley and flywheel on the two ends of the crankshaft; a dial indicator; a special puller or two, although operations using these can usually be done by the machinist. If you work on a front-wheel-drive car with its transmission under the engine, you need a special tool to compress the clutch pressure plate to remove the transmission. Consult the *Clutch* chapter for more information. If you can't remove the clutch, don't attempt to rebuild this engine unless you are prepared to drive the engine to the dealer for him to remove and later to replace it.

A special tool is also necessary to remove most engines: a block and tackle, a chain fall, a come-along, or a cherry picker. The latter is a self-contained hoist on wheels. All these may be rented. Further consideration of these tools is given in the *Tools* chapter. For most rear-engined cars, only a few jacks are necessary to put under the engine to keep it from falling on the ground when it is unbolted from the trans-axle bellhousing.

Removing an Engine

All engines, whether they are in the trunk, next to the driver, or in the front, follow the same general removal procedure. It's nice to have a shop manual to consult. If this is not possible then one of the commonly available repair-manual abridgments is worth looking over, or at least *Chilton's*. The procedures described in these books all originate from the shop manual and are not perfect, so don't be afraid to introduce your own innovations if they look promising.

The first step, often a two-man job, is hood removal. Then, disconnect the ground strap from the battery. When this is done, you can't fry any of the solid-state electronics in the car. Drain the water from the radiator and the block, and the oil from the sump. The plug that is supposed to allow you to drain the block is nearly always plugged. You might be able to open it with a piece of coat hanger. From now on follow an order which is most convenient.

Look at the connectors as you remove them. The type of connector and its position are usually characteristic enough to remind you where the connector belongs. All electrical connections to the engine are disconnected.

Many English cars use direct oil lines for oil pressure and an ether-filled wire-wrapped tube with a bulb at the end for engine temperature. The oil line should be disconnected at the flexible hose, the end with two sizes of hex fittings; one to be held, the other to be turned. The temperature device should be carefully unscrewed from the head or block. If the tube bends it will probably leak and the whole gauge-reading unit and tube will have to be replaced or rebuilt as a unit. This is a large job.

Remove the radiator and hoses, disconnect the exhaust pipe, remove the generator and any other appliances that could get in the way. Also disconnect all parts coming to the transmission (see *Clutch*). If the transmission is to come out as well, as in many small sports cars, leave the starter motor in until after the engine is out. I generally keep all nuts and bolts from this stage in one can.

At this point I shove the car under my chain fall and put it on jack stands. Then, I see if I can knock it off the jack stands before I get under it. Rear-engined cars are jacked in the rear only. On shoebox cars like a Sprite, TR-3, or Datsun 1600, the transmission comes out with the engine, so you disconnect and remove the driveshaft. Remove the intake manifold with the carburetor, if it is easy. Keep the intake manifold bolts separate for convenience. Unbolt the exhaust manifolds keeping the bolts or nuts in a separate container because they are generally special for this application. Leave the distributor body in place for the moment. In a front-wheel-drive car with the transmission under the engine, disconnect the front-axle U-joints and possibly a few other suspension connections as recommended in the book for the car. Conventional engines must be separated from the transmission after the bellhousing bolts are removed. I keep all underneath bolts in a separate can.

There are many ways to do this job. You will be in better shape the greater the number of steps ahead you can see, just as in chess. Don't make moves that aren't reversible (don't break something to get it out) unless the item is to be replaced.

From the top, check to see that everything is disconnected that should be except the engine mounts. Put a sling around the engine and connect it to your hoist. I use two slings. One close to the front and the other, longer one, to the rear. Sometimes hooks are mounted on the engine itself. If this is the case, just connect a single chain between the two hooks, then attach the hoist to the chain. The idea is to get an angle on the engine especially when the transmission is coming along for the ride. The front should be high, the back dropping down at 15 to 30 degrees when it is freely hanging. Before you start to pull the engine, however, return to the engine-mounting bolts.

With the hoist taken up so that there is no slack in the slings, but not too tight, undo the mounting bolts. My practice is to undo one of the single bolts on either side and then unbolt the bracket from the engine on the other. You will see why when you do it. Now begin to draw up the engine. When it is just free of the mounting bolts, pull the engine forward to disconnect it from the transmission in the conventional engine. It can then be pulled straight up, as can the front-wheel-drive engine. More maneuvering is required, if the transmission is coming along. Usually the car will have to be removed from the jack stands. On rear-engine cars, with the bellhousing bolts removed and all other cables, pipes, the wires disconnected, the engine is pulled straight back with a jack underneath the engine to keep it from falling to the ground. I use my transmission jack for this.

As you are removing any type of engine, at this point watch for connections you have failed to disconnect. Make mental notes as to how the procedure might be improved for installation because installation will be harder than removal. On one car, a Volvo P 1800, I had to remove the transmission while the engine was half-way out of the car. That was a pain in the neck. I generally break for lunch at this point.

Engine Disassembly

When the engine is hanging from a hoist and the rest of the car out of the way, remove the clutch assembly. The transmission will have to be removed, of course, if it came out with the engine. Refer to the *Clutch* chapter for details on the clutch and flywheel. The clutch should be overhauled as a part of this job because it is exposed. In front-wheel-drive cars that have the transmission under the engine, the whole unit should be lowered onto a dolly. The screws which hold the transaxle to the block should be removed and the two units separated. The transaxle should be covered and put to one side. (Some machinists prefer to rebuild engines themselves and price their services in such a way that it is cheaper to have them do it. You might check this before you decide to rebuild the engine yourself. In fact, I would say it is common procedure for many garages to buy "short blocks" from a machinist. You might do it this way the first few times, but remember that the greatest accomplishment and fun is in doing as much of it as you can yourself.)

With the flywheel exposed, it can now be removed. This is easier if the oil pan is removed so you can see the crankshaft. Keep the pan bolts together in a can. Notice whether the threads differ on some of the screws, as they might depending on whether they screw into the iron block or into an aluminum timing cover. The handle of a screwdriver can be used to block the crankshaft against the block. This allows the flywheel bolts to be turned. Mark the position of the flywheel on the crankshaft because you want to put the flywheel back in the same place from where it came off. Now pull off the flywheel and put it aside, possibly to be machined if it is grooved or to have the ring gear replaced if it is chewed up.

The engine can now be mounted on an engine stand which makes the work easier. Otherwise the job can be done with the engine lying on its side on a dolly.

Observing the Basic Timing

With the engine mounted on a stand, remove the rocker-arm cover or cam cover. Remove the spark plugs if they haven't been already, to make turning the crankshaft easier. Now turn the crankshaft until the timing marks on the fan pulley and on the timing cover line up at the T.D.C. marks. When they do, the number-one cylinder is at the top of its stroke—either the exhaust stroke or the compression stroke. To determine which, look at the rotor of the distributor or wiggle the rocker arms of the number-one cylinder on a push-rod engine. If the intake valve is loose but the exhaust valve is tight, it is the exhaust stroke, and the crankshaft has to be turned 360 degrees around to T.D.C. again. Now both valves should be loose, indicating that they are closed and the rotor of the distributor should be pointing to place in the distributor cap where the number-one spark-plug wire comes out.

In a flat-head engine the head can be removed to observe the valves directly. On an overhead-cam engine you can observe directly whether both the valves are on the heel of the cam or whether the nose of the exhaust cam is depressing the exhaust valve.

Remove the fan pulley so that the timing cover can be removed. Block the crankshaft so it doesn't turn when the pulley nut is loosened. The pulley itself may be drawn off with a puller—a 7" two-jaw type often suffices. A crankshaft vibration damper may be part of the fan pulley. Check to see if it is cracked or loose. If it is, it will actually aggravate the difficulty it was intended to overcome. Vibration in the crank is a cardinal sin and will wipe it out in no time. The inline six-cylinder engine has the longest crank of those designs generally in use today, so it is most prone to resonant vibration, therefore vibration dampers are usually found on the sixes. There is a seal in the timing cover, which should be replaced.

With the timing cover removed, the timing chain and gears are exposed. Since you positioned the cam and crankshaft for the number-one cylinder compression stroke T.D.C., you should be able to see the timing marks on the chain and gears lined up. This is the position in which you want these parts when you reassemble them. A picture of these positions is usually contained in a shop-manual abridgement.

Now remove the distributor hold-down clamp in preparation to extracting the distributor. Bosch distributors have a mark on the edge of the housing where the rotor should normally point with the number-one piston on T.D.C. compression stroke. Other distributors can be marked with a triangular file. As the distributor is extracted, the rotor will turn 10 to 15 degrees because the gear engaging the cam is spiral cut. Make a second mark on the edge of the distributor housing to indicate where the rotor should be to start the re-engagement of the distributor with the cam. At the same time mark the shoulder of the distributor where it abuts the block. This will show you the normal position of the distributor housing.

Bosch Distributors

Bosch and Ducellier distributors have a two-piece distributor shaft. They are coupled by an offset slot and blade. The distributor in this case pulls directly out without any special problem. It is the bottom section of the shaft which you have to take pains to install in the proper groove. This bottom piece in the Bosch type has a spring in a hole which has to be removed and put into a parts can. A special extractor can be purchased to pull out the bottom section, or a length of 1/4" dowel fits tightly into the Bosch spring hole and can be used to draw out or replace this unit. Observe any special washers that may be at the bottom of this shaft. A picture in a shop-manual abridgement will show a picture of the correct alignment of the offset slot when the bottom section is fully seated.

Engagement of Distributor with the Oil Pump

As if the proper engagement of the distributor gear with the cam were not enough of a problem, you will have to pay attention to distributor shaft/oil-pump engagement. When the distributor is removed, the bottom of the shaft (except for Bosch and Lucas distributors) is a flattened blade. This blade engages a notch at the bottom of the hole into which the distributor shaft fits. The notch is in one of the oil-pump gears. The cam gear therefore turns not only the distributor, but the oil pump as well. When the distributor is replaced, it not only has to spiral into the appropriate teeth of the cam gear, but also to engage the oil pump at the bottom. You can now refer to the *Valve* chapter for procedures to remove the head from the engine—with one major exception: When you remove an overhead-cam head, it is pointless to preserve the timing. Take everything apart keeping in mind that you will have to know how to restore the timing relationships upon assembly.

Cam Followers or Lifters

When the head is removed, we can try to remove the cam followers or lifters. The lifters translate the rotary motion of the cam into an up-and-down movement, which is appropriate to the opening and closing of the valves. The lifters are a machined fit and should be replaced in their original places, if you have a choice. The surface that rides on the cam takes the most wear. It has a thin hardened surface but after some time can pit, get dished, or mushroom. If the latter happens, you cannot remove the lifters from above. The cam will have to come out, so the lifters can drop out from below. If the lifters are anything but smooth, flat, and polished, replace them. New ones are advisable, although reground ones are available. Reground lifters should be hardened again, otherwise they won't last. Valves going out of adjustment fast could be evidence of bad lifters. Sticky lifters produce a tapping noise.

Hydraulic Lifters

Hydraulic lifters used on many American cars must be kept together or replaced as a unit. You can sometimes work on the hydraulic lifters with the head still in place but with the rocker-arm assembly and push-rods removed. On V-8s they can be removed when the intake manifold is off. A special tool is available that can extract them from their recess, if they aren't stuck.

When working properly, they maintain zero valve lash at all engine temperatures, therefore there should be no valve noise. The three causes of hydraulic-lifter malfunction are dirt in the lifter valve, varnish buildup causing sticking, and improper oil level in the sump—either too high or too low. The machinist can help you decide what to do with your lifters.

Camshaft Function

The crankshaft not only turns the rear wheels through the drive train but also turns the camshaft. The cam and crank are either directly geared or are linked by a chain or a belt. The camshaft operates the distributor/oil-pump combination, the fuel pump, and the valves.

The cam gear is larger than the crank gear because it has twice as many teeth; it makes half the number of revolutions that the crankshaft does.

The crankshaft has two basic movements, up and down. The cam determines that one of the upstrokes is compression (both valves closed) and the next is exhaust (exhaust valve open). It also determines that the alternate down strokes are first intake (intake valve open), then explosion (both valves closed). This last is the desired end of what went before; the harnessing of an explosion to do work. Amazing. Thank you, Herr Nikolaus Otto.

In one complete cam revolution, therefore, each cylinder has gone through the four phases of the cycle. The crankshaft, however, has revolved two times. In a four-cylinder engine, each cylinder is going through a different phase of the cycle. If the number-one cylinder is on compression, then the number-four cylinder is on exhaust. And cylinders two and three will be on a down stroke, intake, and explosive phases respectively.

Camshaft Removal

To remove the camshaft, take out the fuel pump, the distributor, and the lifters. If the cam gear is made of a fiber composition, it may require a special puller; leave it for the machinist to do. A flange holds the cam in place. It can be unbolted through holes in the timing gear and the cam can be pulled carefully forward and out.

This process can usually be done with the engine left in the car. The radiator and the grill have to be removed. Sometimes the engine has to be freed from the engine mounts and jacked up a bit also. To imagine the clearance necessary, think of an iron rod, as long as the engine, that must come straight out without interference. If the cam is being removed because the timing gear or chain broke, then a valve job should be done because probably one or more of the valves will be bent. Bending is the result of the piston hitting them as they hung open when they should have been closing. With the head off, the lifters can be removed (you hope).

If the cam gear has to be replaced, the crankshaft timing gear should also be replaced if they directly mesh, as in many four- and six-cylinder engines. If a timing chain is used, it should be checked to see if it is streched too much. A shop-manual abridgement will give you the pertinent data. Spring-loaded chain tensioners should be replaced.

Checking the Cam

The nose or lobe of the cam pushes the valve open. The valve spring keeps the valve closed otherwise. The shape of the cam lobe determines how long the valve is open and how far it opens. The position of the lobe on the camshaft determines when the valve opens and closes. Examine the lobe and the ramp for metal deterioration. Bring the cam with you to the machinist to be examined. Cams can be reground and hardened. As long as the profile of the lobe is maintained, the reduced size can be made up in the valve adjustment. The circular lobe on the cam operates the fuel pump.

Cams usually ride in their metal shell bearings inserted in the block. Their condition and possible replacement is another job for the machinist. The cam doesn't carry nearly the loads that the crankshaft has to carry. Wear on the lobes from the spring-loaded valve train is usually its weak point.

Oil Pump

When the oil pan has been removed, the oil pump is revealed. It is usually held on by two or three bolts. On VWs and Porsches the oil pump runs off the end of the cam and is mounted just below the crankshaft pulley. The oil pump is basically two lobed gears in a housing, which all fit together with very close tolerances. The gears create a suction through a pipe that extends into the oil sump. They then squeeze the oil forward into the oil galleries, which feed oil under pressure to the main bearings, rod bearings, and cam bearings. Oil also goes to the rocker arms and perhaps to a squirter that lubricates the timing gears. In all cars except VWs the oil goes through an oil filter. All cars have a bypass valve, which dumps oil back into the sump if the pressure gets too high.

The bypass valve has a compression spring, which is calibrated to allow opening at a certain pressure. On any car older than 20 years, no matter what the road mileage is, this spring should be replaced because it loses tension over time. You could recover as much as 20 lbs. per square inch of oil pressure by replacing this spring. In a few cars the tension in the spring can be adjusted, but this should not be done to compensate for oil-pump insufficiency. This bypass valve may be a part of the oil-filter housing, part of the oil pump, or in its own drilling in the crankcase.

I generally get a new oil pump when I'm rebuilding an engine. Pump failure may be indicated by badly wiped-out bearings, especially near where the main oil line enters the crankshaft. The oil-pump clearances may be measured with a feeler gauge, however, and if they are acceptable there is no reason to replace the old one. The acceptable clearances will be listed in the engine overhaul data in a shop-manual abridgment. The measurements are made between meshing gear lobes (backlash), between a lobe and the pump body, and between the pump cover and the end of the gears (endplay). If the pump cover is scored, it can be resurfaced by circular rubbing on a piece of 320-grit emery paper placed on a flat surface. While the engine is in the vehicle, an oil pump can generally only be replaced if the oil pan can be removed as well.

Measuring Rod and Main-Bearing Clearance

With the oil pan off we are in the position to measure the health of the engine's muscle, that is, the crankshaft. This measurement and the next few operations can be done with the engine in the vehicle if the oil pan can be removed; crossmembers running under the oil pan may not allow you to do this. On the other hand, there is no need to do this on an engine overhaul if everything is going to the machinist anyway.

Plastigauge is a small calibrated wax wire used to measure bearing clearance. It comes in three ranges; green (.001" to .003"), red (.002" to .006"), and blue (.004" to .009"). The red is used to check for bearings that are too tight, blue is used to check for worn bearings in larger vehicles, and green for checking normal clearances in car engines. To use it, first unscrew and remove a bearing cap.

The bearing itself is a thin half-shell, which will probably come off with the cap. Most of this bearing is steel, with a special thin layer or layers of metal on the inside concavity. The layer that directly contacts the polished crankshaft journal is called Babbit metal. (The formula for this was originally concocted by the inventor Isaac Babbit before the Civil War.) It is fairly soft and rather slippery.

With a clean cloth wipe the journal surface and bearing free of oil. Then place a small piece of wax wire 30 degrees either side of dead center on the journal. The bearing cap with its bearing shell is replaced and retorqued to the factory specification. Remove the bearing cap once again to observe the degree to which the wire was squashed. A gauge comes printed on the plastigauge package, which, when compared to the squashed wire, tells you how much clearance is in the bearing.

The middle main bearing or the big-end (flywheel end) main bearing take the most pounding, so measure one of these. If the clearance is really excessive to the point where oil pressure was a problem on the running engine, then the main bearing bores could be out of round or distorted in some way. The machinist can determine this, and the problem can be remedied by line boring the crankcase.

The rod bearings can be checked for clearance in the same fashion. They might well show more damage than the main bearings. Check your rebuild data, but more than .003"-.004" is too much. When you become an oldtimer, you can try shimming undersize bearings, and then testing for tightness with cigarette paper. Until then let the machinist take care of things for you.

Ridge Reaming

With the head off, we can feel around the top of the cylinder. About 1/4" down from the top is a ridge, more pronounced in some areas than in others. This ridge marks the limit of the wear cut into the cylinder by the top piston ring. The 1/4" above the ridge was the size of the whole cylinder when it was new. If the ridge is very deep, a ridge reamer may have to be used before the piston can be pushed out of the cylinder. Generally this isn't necessary, especially since you aren't intending to save the old rings.

Something has to be done about this ridge, because new top rings won't fit into the old wear pattern and they will shatter or bend the piston land, clamping the second ring and preventing it from funtioning properly. If you are going to rebore the block using oversize pistons, or are going to resleeve the block using standard pistons, the ridge will not be a factor to consider. If you are only going to hone the cylinders and rering the old pistons, the ridge will have to be considered. With honing the reringing you may very well not be taking the engine to the machinist, so three alternatives exist for dealing with the ridge: you may purchase a ridge reamer to take out the ridge, you might chamfer the top ring slightly, or you might get a stepped top ring if you can find one.

Removing Connecting Rods and Pistons

When the connecting rod caps are removed, the connecting rods and pistons can be pushed out by a wooden dowel placed on the backside of the piston with the aid of careful tapping with a hammer. With the first rod and piston removed, replace the rod cap the way it came off. Usually a number on the rod should be the same as, and on the same side as, a number on the rod cap. You will want to put this assembly back the same way, with the numbers on the correct side. Notice whether the numbers face the cam or face away. I mark the front of the rod cap with my marking tool indicating the cylinder from which it was removed.

The piston should be marked with an arrow or a notch to indicate the front. I usually mark them anyway with the number of the cylinder on the front edge of the piston. The reason for all this marking is that the piston-rod assembly is not bilaterally symmetrical. Therefore the piston could be put on the rod backward and the rod could be put on the crank backward.

When all piston-rod assemblies have been removed, the clips that keep the wrist pins in place can be extracted. You can drive the wrist pin out using whatever works for a drift punch. I often use a 6", 3/8" drive extension with a 10-mm socket on the end. I put the piston on a partly opened vise on a bed of rags and drive the pin out far enough to release the rod. I always drive the pin out in such a way that it points to the front of the piston. Some consistent compulsive behaviors like this help remind you of the order of things.

The nature of the problem that caused the engine to be rebuilt determines what to do

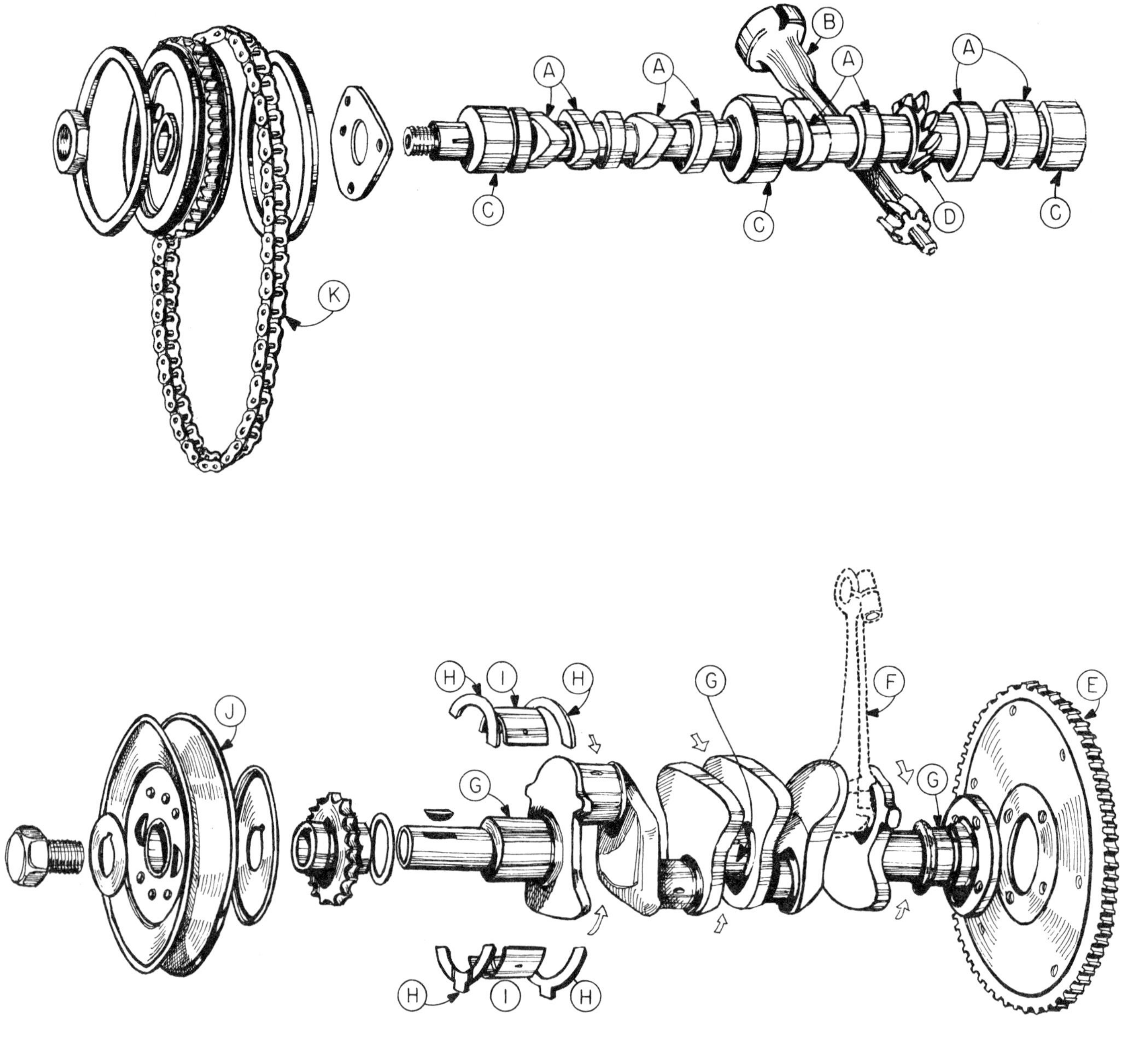

A Valve Cams
B Distributor Shaft
C Cam Bearing Journals
D Distributor Gear
E Flywheel
F Connecting Rod
G Main Bearing Journals
H Thrust Bearings
I Main Bearings
J Crankshaft Pully
K Timing Chain

when the engine is apart. If a rod bearing froze, all rods should be checked and resized. On some cheaper modern engines the rod caps can get stretched significantly under 75,000 miles.

Also check the brass cylindrical bushing in the small end of the rod. The wrist pin goes through this bushing. It has oil holes in it generally, which line up with holes in the rod casting. If these are not lined, the bushing has probably "spun," that is, it is loose and the rod should be rebushed. If the rod is to be rebushed, bring a wrist pin along for the machinist so he can get a correct fit. Rods can be checked for straightness, and should be if significant damage is found in the cylinder or on the crank.

Cylinder Bore

Before examining the pistons check the bore. If there is too much wear, the block will have to be rebored and new pistons and rings installed. If wear is not too great, the bore can be honed and new rings put on the piston.

New bores are circular, but they wear into an oval pattern. The main axis of the oval (the largest dimension) is 90 degrees to the main axis of the engine. The bores also wear in a tapered fashion, with the top ridge demarking the area of greatest wear.

Three methods of measuring bore wear can be used. If you have an inside micrometer you can measure the main axis of the oval at the top of the cylinder just below the ridge and again at the bottom of the cylinder where wear is negligible. Subtract the first measurement from the second. If the difference is more than .015" you should rebore.

If you don't have a micrometer, you can take a *new* standard ring for this engine and slip it inside the piston down to just below the ridge. Measure the ring gap with a feeler gauge. A manual for the particular car might give acceptable limits for ring gap. A rule of thumb for minimum gap would be: .012" gap for a 3" bore, up to .016" for a 4" bore. This gap is to allow for ring expansion as it gets hot. If the gap is in this range or even up to twice those values, you could reuse your pistons and simply ridge ream the cylinders or chamfer the top rings, and hone the cylinders. Different rings are available for your various requirements. The machinist can assist you here.

If you don't happen to have a new standard ring lying around, you could take the whole problem to the machinist. He will probably recommend reboring to be on the conservative but expensive side.

Another type of cylinder wear leaves vertical scratches. If they are too deep, oil gets pumped by into the combustion chamber. Your fingernails are sensitive for detecting this damage. Grooves should be removed by reboring. The machinist will have to determine how much will have to be removed to clean up the bores.

Since many modern cheap engines need engine overhauls at less than 100,000 miles, the possibility of reusing your pistons and rings is greater. After you have driven 150,000 miles, you need reboring and therefore, oversize pistons and rings. The Vega aluminum block was never intended to be reused. You were supposed to get a new one, if the old one showed wear beyond the normal limits. However, due to the ingeniousness of the rebuilding industry, this engine block can be bored out and steel sleeves inserted, making the engine potentially better than new. On some cars the cylinders are removable sleeves, which are replaced as a unit with the cylinders and rings. In this case you are always using standard sizes.

Pistons

If you decide to reuse your pistons, they should be checked for wear, and if they pass the test, cleaned. Find the piston with the loosest top ring. Try to insert a .005" feeler gauge between the land and the side of the ring. If it fits, junk the pistons, unless the cost of parts is very high (Mercedes, for example), in which case you can regroove the pistons and have steel inserts put in to restore the groove to standard size. This service may not be readily available in all areas.

Next, clean the carbon off the top ridge in a cylinder and slip a piston into the top an inch or so. Measure the clearance between the piston and the cylinder. Normal clearance is .002" for smaller pistons (2"), to .004" for larger ones (4"). If clearance is more than .003" in excess of these values, the piston skirt is collapsed. This can be rectified by knurling, which is a process of raising the metal on the piston skirt. The process tends to prolong the life of pistons and bores by improving lubrication.

If the piston passes these tests, it should be thoroughly cleaned. The top of the piston should have all deposits removed with a scraper. The ring grooves can be cleaned with broken pieces of old rings. Remove all varnish deposits down to the metal. The new rings, being unworn, will be too tight in the cylinder, if these deposits are left. Glass-bead blasting will clean the pistons nicely if you wish the machinist to do it. Sand blasting removes metal and is therefore unsuitable.

Most of the discussed repairs can be done with the engine still in the car, if the oil pan can be removed. Some so-called complete engine overhauls are done with the engine still in the vehicle. If the crankshaft is ignored in an overhaul, the car could run very nicely at *first*, since performance depends mainly on good rings, cylinders, and valves, that is on good compression.

Crankshaft and Block

You can renew the main bearings with the engine in the car, if the oil pan can be removed and the crank doesn't need to be ground. For many cars you can obtain new .001" to .002" undersize rod-and main-bearings. If the journals are worn no more than .003" to .004" and are clean and polished, it would be acceptable simply to use undersize bearings without machining. Check the clearance, using Plastigauge. The main-bearing half shell in the block can be coaxed out by turning the crankshaft, and with a gentle push with a small screwdriver on the side that doesn't have the recess punched in it. Be careful not to scratch the journal. The new one can be

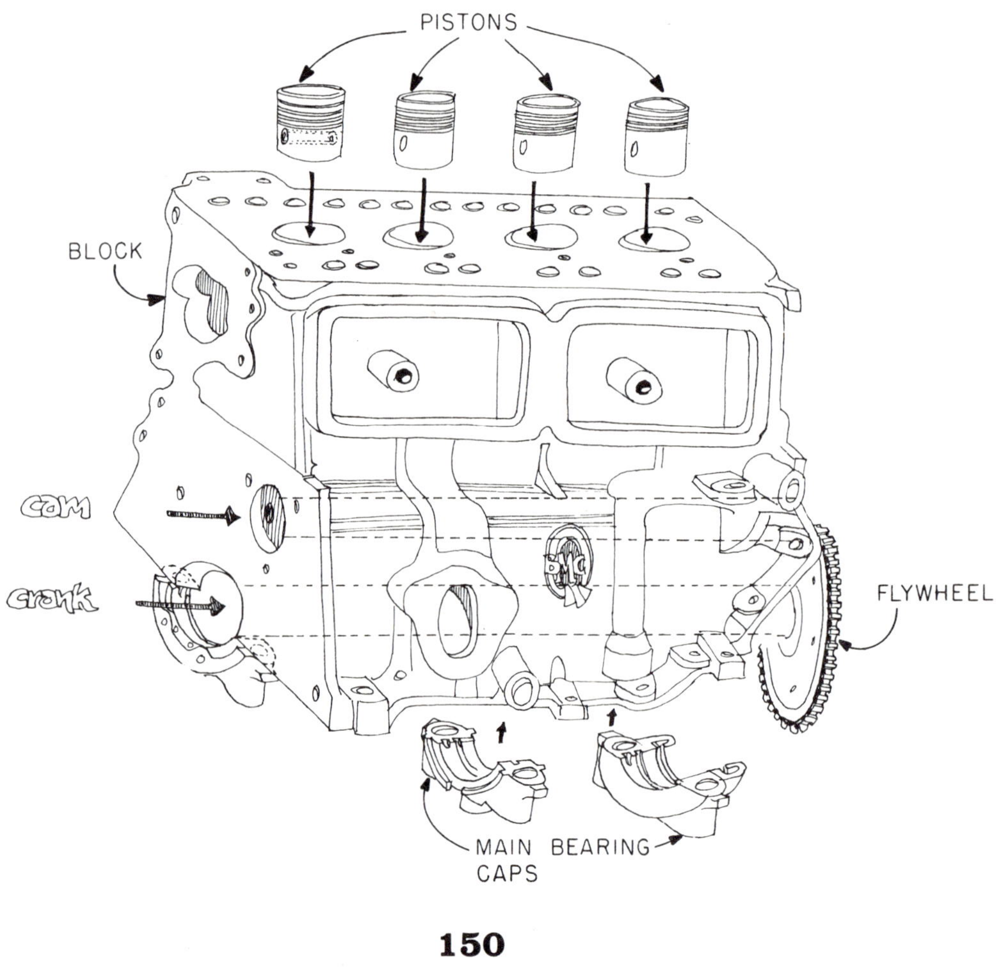

oiled and slipped back in the same way. If there is excessive journal wear, the engine has to come out.

With the engine out, the crankshaft main caps can be removed. Number the caps if they aren't already. Put the number on the front of the cap, so you can put it back exactly the same way. With everything apart, look at all the sliding and rubbing surfaces. They should be polished and true; no pitting, dishing, or grooving. If the wear on the crankshaft is excessive, it should be checked for cracks. If there were some seized bearings or excessive bearing clearances, ask the machinist to check the crankshaft for straightness.

You can quickly test the crankshaft for cracks by holding it by one end and striking it with a hammer. If it rings clear like a bell it's OK. If it sounds a muddy dead note it is probably cracked. The machinist can magnaflux it to be sure. Cracks are mostly not visible to the unaided eye.

In fact, a fair amount of damage in the engine can only be seen if the light is just right. Direct light often obscures metal wear, and only when the light is off to one side creating a shadow can some grooving be seen.

Many machine shops have a truck that can pick up the block and return it for you; whether they will is another story. It's worth a try. Until you know more about what you are doing, bring all the main engine parts to the machinist and get him to go over them with you. This examination includes the bearing shells, so he can see the type of damage incurred. Have the block boiled, which is a method of cleaning it. Ask the machinist for a deadline and call him on it, so he knows you're serious, but until you get to know how he works assume it will take twice as long.

If you want a really nice job on an engine, you can have the fan pulley, crank, rods, pistons, and clutch all balanced as a unit. This is only done at special machine shops. You could end up with an engine that will last longer than the one originally installed in the factory. Complete engine balancing is expensive and hardly ever done. If a large number of nuts and bolts holding things onto the engine, as well as those holding the engine together, were loose, consider having the crank balanced unless you can explain the looseness in some other way, that is, a few very weak cylinders. I have seen vehicles that came from the factory with a hard vibration at idle, which must have been due to a drive-train imbalance. Some balancing you can do yourself. You can weigh the connecting rods and grind metal off the heavier ones until they match the lightest rod in weight. The part to grind on the rod is the section that already appears to be machined. You can also match the weights of the pistons by removing metal from inside the piston in the skirt area. You might check the exhaust manifold for warping. If it is warped, the machinist might be able to resurface it. I always have VW and Porsche crankcases checked for cracks. In fact, I go to a machine shop that specializes in VW and Porsche when I am working on these engines.

Speaking of cracks in the block, epoxy can be used to seal cracks in cast-iron blocks as long as the crack isn't in the cylinder area: use epoxy on the bottom half of an eight-cylinder block and or all of the pushrod side of a six-cylinder. Clean up the crack with a grinder down to bright metal. A long crack should have holes drilled at either end to keep the crack from extending. Make the holes just the right size for a couple of self-tapping screws to be screwed and epoxied in place. Welding is required for cracks higher up on the block—too hot for epoxy.

Freeze plugs in the block or head can be replaced by the machinist. If he is doing his job, the freeze plugs would be included in this regular inspection. You can replace the plugs yourself with a special rubber plug which fits very tightly in the plug hole. The old plug has to be removed completely, the sealing edge must be clean and *dry*. Then the rubber plug can be put in. Measure the hole accurately before you go to the store.

When you have a clean block it is nice to refinish with an emery board the machined surfaces such as the top of the block, and the places where the timing cover and oil-pan mount. Take a 16" piece of a two-by-four and staple a strip of 320 emery paper to it, available at your neighborhood parts store. Use it to clean up the flat machine surfaces, then blow the block clean.

Putting the Engine Together

With clean, machined or new parts back from the machinist, work scrupulously clean. Have plenty of clean rags and solvent around. From now until the head is installed, don't answer the phone or flirt with the neighbor. Notice the bearing size marked on the box. Besides the part number it will say STD (standard), .010, .020, or .030. If the crankshaft was just polished, it will be standard, otherwise it will be marked the amount that was taken off. The bearings will be a corresponding size larger. If the crankcase was linebored, the outside dimension of the bearing will also be larger.

With your first engine overhaul, I recommend mounting a rod to the crankshaft outside the block (normal procedure of VW and Porsche). This will give you an idea how they should fit. When the rods are torqued on, you should be able to pick them up and let them drop slowly of their own weight. If they hang up, take things apart and clean them up. Polish the journals with 320 emery. Don't do anything to the bearing shells. If these fall too fast, Plastigauge them to see if the oil clearance is too great. If you are experienced with the feel of the rods, take them off and proceed to the installation of the crankshaft into the block.

Now install the bearing half-shells in the block. They are indented, so they will fit properly only one way. The bearing may have a hole that lines up with a dowel in the block. One of the bearings has a collar or there are two semicircular flat pieces called thrust washers, which are bearing surfaces. These determine the amount of front-to-back movement of the crank, called endplay. This will be measured later. If you have trouble remembering where the bearing shells go, refer to an abridged shop manual for a blow-apart picture, which will show the correct position.

During the first short period of initial starting after the engine is together, it won't be oiling itself until oil pressure builds up. Therefore lubrication has to be put on these gearing shells. Every mechanic has his own trick. Some people like Lubriplate, which is also used in the brakes. You can buy special expensive assembly lubrication that looks like burn ointment. A mechanic I know who rebuilds engines for truck fleets just uses ordinary 30-weight oil. Take your pick.

Put the crankshaft into its bed of oiled bearing half-shells. The rear of the crankshaft has a hole for the nose of the transmission imput shaft. The rear of the block is recognized by the bellhousing screw holes. See if the crank turns by hand—it should. If it doesn't turn, the bearings are too thick. Check the bearings themselves, they are marked on the back. I once was given oversized bearings in a box marked standard. Install the main bearing caps torquing them to their recommended values. Put the fan-pully nut in temporarily, so you can turn the crankshaft with a wrench. Turn the crankshaft to see how much drag there is. It might take a fair pull to break it free the first time. Tap the bearing caps a bit to encourage the bearings to seat. If the shaft is frozen or very hard to turn, undo the caps and clean everything up again. Take a long strip of 320 emery cloth and polish the journals. If any one of the bearings was doweled in place, then possibly the bearing wasn't aligned properly and it was "pinned" on installation. If this happened, you will see a new hole in the bearing and you will have to get a new bearing set. On occasion you will get an imperfect bearing. You can Plastigauge the bearing as mentioned earlier. If none of these things help, perhaps the machinist should put this engine together.

If the crankshaft is turning freely, the bearing cap bolts should be locked. Sometimes locktabs are used, sometimes wires. I use Locktite on the bolts instead of lock tabs or wires.

Measure the endplay of the crank. With a screwdriver wedge the crank to one end of the engine or the other. Take a feeler gauge and measure the clearance between the side-thrust main bearing and the crankcase. The space is on the side opposite the direction in which you forced the crank. The allowable play could be anywhere from .004" to .010" or more. Check your manual. Not enough play, and the crank will jam when it's hot. Too much, and it will pound the crankcase on clutch engagement and eventually cause connecting-rod problems.

Wet-Sleeve Cylinders

If you are overhauling a block that uses wet sleeve cylinders, keep the pistons together with the cylinders in which they came. Arbitrarily mark the side of each cylinder with a number (one to four for a four-cylinder engine). Then mark each enclosed piston with the same number that is on the cylinder. Now remove the pistons from cylinders and prepare to install the cylinders in the block, each cylinder going in its correct number hole, that is the number one cylinder going where the number-one spark plug will be. The pistons will go with their appropriately numbered connecting rod in a later step.

The cylinders may be marked with a dab of paint, which indicates the correct orientation of the cylinder in the block. The paint dab should point toward the number-one cylinder. If the cylinders have rubber or neoprene seals leave them aside until the next step is completed. Be sure the shoulder or shoulders in the block against which the cylinders abut are clean. If any type of metal ring is to be installed, it is both for sealing purposes and for shimming the cylinder, which is supposed to protrude above the surface of the block from .004" to .009" depending on the manufacturer. This slight protrusion is necessary to insure a good seal with the head gasket and head. Install the metal seal and the cylinder in the block. Take a straight edge, place it across the mouth of the cylinder, and measure the clearance between the straight edge and the surface of the block with a feeler gauge. In a Peugeot-type engine, install the cylinder (without its rubber seal) directly and make the measurement.

If the measured space is too small, you will need a thicker metal ring from the dealer. If the space is too large, a thinner ring is called for. For a Peugeot, paper rings, which fit at the top of the cylinder, are available at the dealer in different thicknesses. In most other types, remove the cylinder and measure the thickness of the metal ring with a vernier caliper—figure out what you must add or subtract from that to get a ring of the appropriate thickness. If you are using old rings to make the estimate, be sure to measure the crushed part of the ring. (A used copper ring will be crushed around .002" from new thickness.) Do this for each cylinder separately.

In a foreign car, convert your values to metric measurements and take a trip to the dealer. Bring your vernier caliper because you may have to do some measuring when you get there.

When you have the correct rings to achieve the desired protrusion above the block, the rubber seals (if any are used) may be installed on the cylinders. The cylinders may now be inserted into the block. You may have to do some gentle tapping with a plastic-head hammer. In a Subaru wet-sleeve-cylinder engine the head should be installed with the cylinders in a vertical position because they have no support in the top of the block and could be misaligned if done horizontally.

Pistons to Rods

Take the cleaned or new pistons, wrist pins, and rods out into the sun or warm them up somehow. The wrist pin is a very tight fit, but should neatly push into place if warm. There are two ways to put the piston on the rod. Rarely will you find an engine where the pistons can go on either way and be correct. There is always some code to finding which way is correct, if you failed to mark the pistons adequately on disassembly. Refer to your repair-manual abridgement. Tap the wrist pins in place and put in the snap-ring retainers.

Piston Ringing

You have clean and checked pistons, and new rings. The top ring holds compression, the middle ring has a mixed function, and the bottom ring is an oil scraper. The oil scraper keeps oil from being pumped up into the combustion chamber, although a little oil should get past to lubricate the top ring. Put a new top ring into one of the cylinders and check the gap (as described in the *Cylinder Bore* section) just to see if they are the correct size. The abutting ends can be filed to increase the gap.

There are ring expanders, which open the rings up so they can be slipped down over the piston. I just work the rings on carefully by hand. They are very brittle. If one breaks, you will have to get a new set unless you have a good connection with a parts store and can go to a place like Seal Power where you can buy rings singly. The rings come with instructions. There is usually a mark on the side to indicate which side is "up" if it matters. The gaps in the rings should be staggered equally around the piston, so that no gap lines up with any other.

Piston to Crankshaft

Take a piston/rod combination and imagine how it is going to go into its cylinder, through the top. Notice where the number or forge marks are on the rod, and the mark on the front of the piston. Are they all pointing in the correct direction? If the piston is pointing forward and the rod mark is on the wrong side, you have the rod reversed on the piston. Correct the situation.

Some people put two pieces of rubber hose over the rod studs to avoid scoring the crankshaft journal. Notice the box numbers for the rod-bearing half-shells to see if you have the correct size. Insert them in the rod, and the cap in their appropriate position. Oil the bearing shells.

The rings have to be compressed into their grooves to enable the piston to fit into the cylinder. I prefer the type that has a metal band clamped around the piston by a locking plier (K-D tool 850). With the rings clamped tightly and held against the block with one hand, tap the end of the piston with the handle end of a hammer. If it won't go with mild force, either a ring has popped out or the rod is hanging up on the crank or the crankcase. You will have to guide the rod through the crankcase and around the crank journal.

With the rod bearing in place and lubricated, put on the rod cap from the other end making sure the numbers or marks are both on the same side. Locktite the nuts and torque them down. With each cylinder installed turn the crank. The rings introduce a fair amount of resistance, however, you should be able to turn the crank without breaking your back. If you make sure everything turns and moves as you install it, you won't have anxious second thoughts just as you are ready to start the engine. It's too late then.

Cam Installation

The cam can be installed when the journals have been lubricated with whatever your assembly lubrication is. The bearings were checked and possibly replaced by your machinist. A plate with a new gasket may go on first in some engines.

Line up the timing marks and install the timing gears or chain as shown in the book or according to the marks you yourself put there. When this is done, the number-one cylinder should be right up top. In some overhead-cam engines only the bottom chain will be installed now along with idler gears and the chain tensioner. If only one chain or belt is used, it will be installed after the head is in place. Replace the oil-squirting nozzle, if one is used, so it is pointing to where the timing gears mesh.

Replacing the Head

Place the engine upright and install the cam followers or lifters in their original places. Lubricate them with your assembly oil. They should drop smoothly into their bores. If they don't, try polishing the lifters with a piece of 320 emery paper.

Now install a new head gasket. Steel or copper gaskets should be painted with your favorite sealer. If the gasket has bends in it, they can be carefully pounded out with a soft-headed mallet. Rubber-asbestos gaskets need no sealer. Usually the numbers on the gasket face up, or it will say up, top, or "oben." Put on the clean, rebuilt, resurfaced head, torquing the head bolts according to the pattern suggested for your engine (for more details refer to the *Valve* chapter.

Install the push rods and the rocker-arm assembly on a push-rod engine. With an overhead-cam engine install the cam timing gear and chain, paying attention to the timing marks. With a double overhead-cam engine you may be able to find a metal guide from a friendly specialist in the particular car, which will fit over the nose of the exhaust cam and the intake cam for the number-one cylinder. This guide will exactly fix the position of the two cams in the place they want to be when the number-one piston is T.D.C. compression stroke. For the sake of clarity, keep the number-one piston on T.D.C. compression stroke.

Basic Valve Timing

Let's suppose you have the crank, pistons, cam, and head installed and you find two conflicting sets of timing marks, or none at all, or you are installing a cam with a different profile. This means going back to basic valve timing. In *Chilton's* tune-up section or in the shop manual for the car you can find how many degrees before T.D.C. the intake valve opens. It is unique for each engine, so be sure you are looking at the right listing. It will range from 7 to 65 degrees.

To know what "before T.D.C." is, you have to know which way the crank normally turns in this engine. Temporarily install the distributor in any position. Look at *Chilton's* tune-up section for the picture of the distributor and notice which direction the rotor is supposed to turn—clockwise (c) or counterclockwise (cc). Now go back to the engine and turn the cam in both directions. Notice which way you have to turn it to make the rotor go in the desired rotation. Remembering this direction, observe how the cam and crank are hooked together. If they are linked by a chain, the cam and crank will turn in the same direction. If they are directly geared, they will turn in opposite directions. Now you know in which direction the crank turns and therefore what "before T.D.C." means.

Turn the crank so that the number-one piston is roughly half-way down the cylinder. Now insert a wooden dowel into the spark-plug hole (be sure you can get it out). Now gently turn the crank so that the piston slowly eases up the cylinder until the dowel blocks it near the top of the cylinder. Mark a tooth on the timing gear of the crank, and just adjacent to that scribe a mark on the crankcase. Now holding the dowel in place, turn the crank around the other way until it comes up to stop against the dowel again. Scribe the crankcase opposite the marked tooth on the timing gear. Remove the dowel. Now measure half-way between the two scribes on the crankcase, and the point is T.D.C.

Now mount a timing wheel on the end of the crank. The endpaper of some large *Chilton* volumes has a timing wheel printed on it. Otherwise they can be found at some automotive-parts stores. Turn the crank until you think the number-one piston is right at the top of the cylinder. This will probably be at the point where the key slot in the crankshaft is centered opposite the scribe mark on the crankcase. Fix the timing wheel so that zero degrees is just opposite, and on a direct line with, the scribe on the crankcase. Now turn the crank so that the piston moves down, then repeat the dowel trick noting how many degrees short of T.D.C. the dowel blocks the movement of the piston, first clockwise then counterclockwise. The number should be the same. If it is not, the timing wheel should be rotated one-half the number of degrees of difference between the two in the direction of the smaller number. Try it again until the timing wheel is blocked the same number of degrees either way the crank is turned.

Turn the camshaft until you can observe that both valves are closed. Adjust the valve lash. Now mount a dial indicator on the head and place the plunger on the end of intake-valve number one so that it will measure any movement in the valve. Turning the cam in the correct direction (remember the direction the distributor rotor turns) notice the moment you begin to get movement or opening in the valve. Stop! Now turn the camshaft before T.D.C. the number of degrees recommended for the intake valve to open. Use the timing wheel and the T.D. C. scribe on the crankcase as your guide. The crankshaft and camshaft should be geared or chained together in these positions. The procedure described may not be the best for your particular problem. For example, the spark-plug hole might be right on top of the combustion chamber. In this case it might be easier to measure T.D.C. with a dial indicator sticking down into the spark-plug hole. At any rate, it is the basic principle that is important. For more information consult *Auto Engines and Electrical Systems* by Blanchard and Kitchen.

Timing Cover

Before the timing cover is installed, the dished washer should be put on the end crankshaft (if there is one) concavity facing away from the engine. The function of this washer is to deflect oil away from the opening in the timing cover. Next, tap the fan-pulley key into its slot. Now you can install the timing cover with its new seal and gasket. Don't tighten down the cover bolts until the fan pulley is installed so that the fan pulley can center the oil seal and timing cover; then tighten them down.

Distributor Installation

When the fan pulley has been replaced on the front of the crankshaft, the pointer on the timing cover and the T.D.C. notch on the fan pulley can be lined up. If the timing marks are on the flywheel and bellhousing rather than on the fan pulley and timing cover, consider installing whatever rear-main-seal parts are necessary and the flywheel. Once the timing marks are lined up, the number-one cylinder is at the top of the block either on compression or exhaust. On compression, the two valves for the number-one cylinder are closed (up); on exhaust, the exhaust valve will be open (down).

To determine which way the crankshaft is supposed to rotate, put a large wrench on the pulley nut and turn it by hand observing the movement of the number-one cylinder exhaust and intake valves. Watch particularly the intake valve. As it comes up (closes) does the exhaust valve start to go down? If so, you are rotating the crank in the wrong direction. If, when the intake comes up, there is a pause of more than a quarter turn before the exhaust valve begins to move, you are rotating the crank in the proper direction. The pause is the compression stroke, which should follow the closing of the intake valve.

Now that you are turning the crank in the right direction, stop during the compression stroke with the timing marks lined up. Now insert the distributor so that when it is fully installed the rotor points approximately to the number-one cylinder. Consult the marks you made on the distributor as you removed it or look at pictures in a manual, to see what alignment is the proper one. When you have what you think is the correct position, try installing the distributor in the cam-gear groove just before and just aft of that position to confirm your guess. Don't rotate the distributor body to match the rotor. The distributor cap should be positioned as shown in *Chilton's* tune-up section. Make the rotor suit the normal position of the distributor body. The car won't start if you don't get this right, so take the time to avoid the error. Don't forget to replace the rubber O-ring around the distributor housing if there is one.

Now with the distributor in place, rotate it so that you can see the points just beginning to open. Clamp it. This is an approximate timing, which will be good enough to start the engine. The rotor will actually appear to be nearly past the number-one spark terminal in the cap. You can adjust the valves at this point, doing each pair when the rotor is pointing at or near the cylinder's distributor-cap terminal. As a check to see if everything is as it should be, turn the crankshaft around and notice a "woosh" out of each spark plug hole on the compression stroke. No woosh? You either have hydraulic lifters (cam followers) that need to be loosened a bit or something is drastically wrong. Backtrack until you find it. Without compression, the engine won't start.

Bosch Distributors

Bosch distributors have a two-piece shaft, therefore the bottom section must go in first and requires special instructions. Two washers may precede the bottom section of the shaft. These go in a recess at the bottom of the hole. Oil these washers and slip them onto a long skinny screwdriver. Hold them near the handle until the blade of the screw-driver is inserted all the way down into the hole. The screwdriver acts as a guide as you let the washers drop into their recess. Use the screwdriver to center them in their correct position. You don't want to have to fish these washers out, if they fall into the oil sump.

Now the shaft is inserted in the hole. It has to be oriented so that the off-center slot on top is at a particular angle. Refer to your shop-manual abridgment for a picture. A spring hole in the center of the shaft makes a nice tight fit for a 6" piece of 1/4" dowel that can be used to raise and lower the shaft into position. With a little experience you will see that to get the offset slot in its proper alignment, you have to start the engagement of the spiral gear with the gear on the camshaft some 10 degrees either clockwise or counterclockwise of the final desired position. The shaft turns as it seats into the bottom of the hole. Don't try to force engagement of the gears. When you think you are in the right position, experiment by removing the shaft and sliding it into engagement with the grooves just before and aft of it to confirm the correctness of the position. Once in the correct position, hold the shaft in place as you remove the dowel and put in the spring. The distributor now goes in the only way it can. Turn the rotor while pressing it down until the off-center tongue engages its groove. This description adequately covers Bosch distributors used on VW and Porsche engines where the oil pump is not driven by the distributor shaft.

It is not uncommon to get the distributor in 180 degrees out of phase, especially on Bosch distributors. This means that the rotor is pointing to the number-one plug wire on the exhaust stroke. All one has to do is move each plug wire to the opposite position on the distributor cap, and everything will work properly. Of course this may be confusing for the next mechanic who looks at this car; and you may be the next mechanic.

Oil Pump/Distributor Engagement

With the distributor in place, you can install the oil pump. Disassemble it and remove the pump gears. Install the body of the pump around the distributor shaft noting the correct placement of any rubber seals, often standard O-rings, between the pump body and the oil-passage drilling in the block. Oil the pump gears and install them in the pump body paying special attention to the engagement of the distributor shaft with the appropriate gear. Put the rest of the pump together.

If you installed the oil pump and the oil pan before you put in the distributor, you will have to fiddle with the engagement of the distributor and the oil-pump gear. The oil-pump gear will have to be aligned with the aid of a long skinny screwdriver so that it accepts the end of the distributor shaft which, in turn, must be installed in just the right groove of the cam gear. You achieved a proper engagement if the distributor rotor is pointing to the correct place, the distributor body is aligned as shown in *Chilton's*, and the distributor body is all the way down and flush with the block.

Reinstall the cleaned oil pan. If it has a bad dent in it, consider finding a better one at the junk yard because a distorted pan is hard to put on and will invariably leak. If the bolt holes are flared, use a hammer and anvil to flatten them. Mount the oil pan with a new gasket and non-hardening gasket cement on the pan side of the gasket. When replacing the screws, notice if some of these go into iron and some into aluminum. The screws which go into aluminum have a coarser thread. Then tighten them gradually in two or three turns around the pan.

The Rest

Now is probably a good time to replace the flywheel. Resling the engine from your hoist and unbolt it from the engine stand. Install any rear-main-seal gaskets and housings that may be necessary and remount the flywheel in its original position. My machinist says it's insignificant if it goes on differently, but, as far as I'm concerned, good shop practice is to put things back together as nearly the same as the original as possible. Use Locktite on the special flywheel bolts and torque them down to the suggested value. To keep the crankshaft from turning, attach a breaker bar and appropriate socket to the pulley nut in the front of the engine and have someone hold it or block it against the ground. Attach a new clutch assembly to the flywheel (see *Clutch*). With a front-wheel-drive car, which has the transaxle under the engine, lower the engine with a hoist carefully onto the transaxle using a new gasket and a liberal amount of silicon sealer. The clutch assembly can now be installed by the special procedure required for this engine.

Putting the engine back will take longer than removing it. The most difficult installations are usually in the smaller cars where the transmission and engine are kept together as a unit. Actually the most important part of the job is getting the engine safely slung from the hoist and at the correct angle. The best angle is determined from the experience gained from removing the engine.

After the engine is in place many reconnections need to be made: gas lines, water lines, exhaust manifold (there may or may not be a gasket for this), electrical and mechanical (clutch, gas pedal, connections). Put oil in the engine. As a precaution look up the correct amount of oil that should be put in this engine, then check the oil dipstick to see if it reads near full when that amount is put in. Recheck the oil level after the engine has been started and then stopped once. Re-mark the oil dipstick if it is not reading correctly. Put water in the radiator. If the car has been sitting for a while before the engine was rebuilt, put a clean gas filter in the gas line. You might even drain the gas tank and put in fresh gas.

The first ten minutes of running time on a rebuilt engine are critical. As in major surgery, the vital signs should be closely watched. As you start the engine it may take a while before the carburetor is reprimed with fuel. Watch the oil-pressure gauge or lights. In a tight engine you should probably be reading normal oil pressure before the engine starts, but if not, at least after it kicks over.

I always put new plugs, points, condenser, rotor, cap, and often plug wires on a rebuilt engine to eliminate all weaknesses in this problematical area. Watch the water temperature once the engine is running. A fair amount of smoke will come off the heating engine for three or four minutes when oil burns off the exhaust manifold and its new gasket. Look underneath for water or oil leaks. You may have missed a hose clamp or oil fitting.

You are now experiencing either one of the greatest thrills or one of the heaviest disappointments of a mechanic. The car may have to be tinkered with in a week or so after this—usually small things like retiming or retightening some screws. The head bolts should be retorqued after the car has used a tank of gas. The cooling system may "burp" up a steam bubble after an hour or two of running time. Be sure to top up the radiator. New hydraulic lifters may flutter when the engine is under a load for a short time until they work in.

Chapter 16:

Standard Transmission and Differential

We make our destinies by our choice of gods.

—Virgil (70-19 B.C.)

The engine can run only in a fairly narrow range or r.p.m. considering the load it has to push and the speeds at which we wish it pushed. The transmission allows us to select three or four different ratios or mechanical advantages, whatever is suitable for a given load and road speed. The effective torque range for the average engine is roughly 1,000 to 4,000 r.p.m. The engine speed at idle is the speed required for the engine to overcome its internal resistance and run its appliances (fuel pump, etc.). The clutch can only be fully engaged at around 1,000 r.p.m. if you are to avoid stalling.

Not much attention is required by a transmission except to check the fluid level and change it occasionally. When I buy a used car I always change the transmission fluid because it probably has not been done since the car was new. 90-weight oil is normally used in transmissions, but there are a few exceptions. On an old Ford truck I once owned I used 140-weight oil to ease gear engagement. I had to double-clutch on the first shift or two on a cold morning, but after that things ran smoothly. I would not recommend 140-weight oil on newer transmissions. I do recommend mixing in a molybdenum disulphide transmission additive with your regular lubrication.

Column Linkage

Many shifting difficulties are actually linkage problems. The column-shift linkage can be checked and adjusted in the following way: Disconnect the shifting arms from the side of the transmission. With a little effort you should be able to move the two arms on the side of the transmission. They each have three positions. An interlock permits them to move only if the other member of the pair is in the middle position. If the arms are impossible to move, something is jammed inside and it's time to take out the transmission.

After checking the shifting rods for free movement, set them each in the neutral or middle position and replace the linkage rods. Go up top and look under the hood at the steering column just at the firewall. The shift-linkage rods attach to levers that are activated by either a finger, or a slotted blade that engages pins on the shift levers. The finger or slotted blade is attached to a shaft that is moved by the shift lever inside the car. The finger normally rests in the groove of the linkage farthest away from the firewall. This is the second-to-third shifter. Have somebody move the shift lever in the car between second and third to see if the lever at the transmission is being moved fully into second and third. The neutral position should be in the middle of the range of movement. It can be adjusted by lengthening or shortening the arm running from the steering column down to the transmission. This is managed by removing one end of the arm, loosening a lock nut, and screwing in or out a threaded connector.

Now that the second-to-third linkage is established in its proper position, the first-to-reverse linkage should be lined up. It should be just opposite, so the finger engages its slot directly upon somebody lifting the shift lever toward them inside the car. If this is not the case, the shift finger will be blocked and shifting between first and reverse will be impossible. The position of this linkage is also determined by the shift linkage connecting the steering column with the transmission.

Actually shifting from first to second across the H pattern may be facilitated if the finger-engagement notches are slightly staggered. Experiment will show you the best position. Sometimes the two shift levers which are engaged by the shifting finger get spread, making the shift pattern sloppy. If so, bang them together with a hammer.

Floor Shift

The linkage on a floor-shifting transmission will probably give less trouble than column shift linkage unless it is a conversion from a column shift, in which case it may very well give more trouble. Conversion kits, besides being frequently rather difficult to reach, tend to change their position. When they shift position they don't shift gears. Shift linkages should be case hardened so they will not bend. Cheap conversion kits often are not case hardened.

Transmissions that are designed to be top shifting instead of side shifting as in column-shift transmissions, are not completely trouble free as far as linkage is concerned. The Volvo floor shift, for example, has a brass fitting that wears out making the H pattern sloppy. When the shift pattern is sloppy on any transmission, it will tend to be destroyed sooner. This brass fitting, which is part of the top plate, should be replaced when a Volvo transmission is rebuilt.

There are detent springs, which press ball

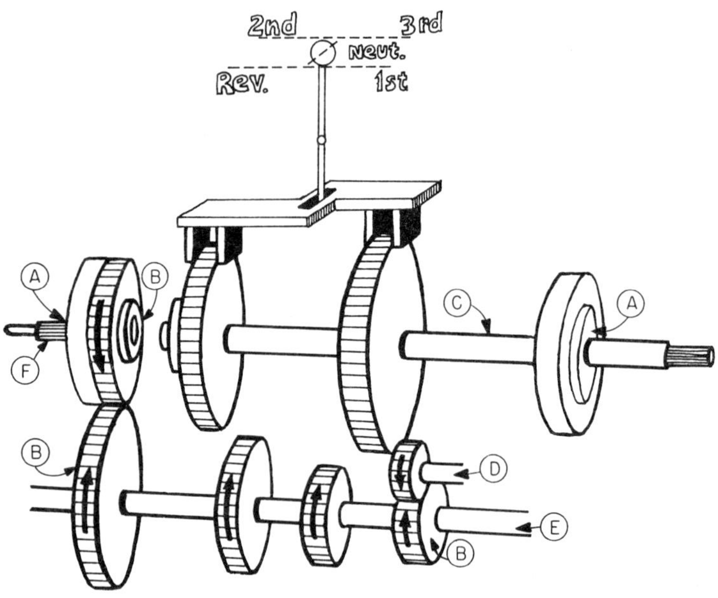

A Ball Bearings
B Roller Bearings
C Main Shaft
D Reverse Gear
E Counter Gears (Lay Shaft)
F Input Shaft

bearings into notches cut in the shift rods of most floor-shift transmissions. The function of these is to keep the gear in position after it has been selected. On older cars the spring will have lost tension. Popping out of gear under load can be due to weak detent springs. This can be remedied with a new spring or by shimming the old one. Be careful not to make it so tight that it won't shift at all.

Sometimes the screw holding the shift fork on the selector shaft gets loose, and the shift lever just flops back and forth without causing a shift. This would probably only happen between two gears. This problem, as well as the detent spring, can usually be remedied without removing the transmission. Put Locktite on any loose screws.

Otherwise, crunching and grinding followed by the inability to shift or with no power transmitted to the rear wheels, suggests transmission problems. Jumping out of gear, or ticking, also suggests transmission problems. Ticking in first, due to a chipped tooth, can be lived with, but drain the transmission fluid to get the loose metal out. When a second tooth goes, it is probably only a matter of time before the gear will strip and no longer transmit power. A bad clutch is usually not noisy, it just slips and burns. A bad driveshaft is quickly checked visually (see *Clutch* chapter). A bad rear end may sound like a bad transmission but is identified by the source of the sound. Have somebody walk along the car as it rolls down the road making the sound. A rear axle that has stripped a drum will also be localized by the origin of the sound. Refer to the *Clutch* chapter for transmission removal.

On front-wheel-drive vehicles the engine has to be removed to get at the transmission (see *Engine Overhaul II*). It is actually a transaxle. Transmission gears and the differential gears are in the same housing. This type of vehicle also has a transfer case, if the transaxle is under the engine. The transfer case connects the clutch shaft by gears to the input shaft of the transmission mounted below it. The clutch has to be removed before the transaxle can be separated from the engine. Special tools are required for relieving the tension on the clutch pressure plate. As mentioned earlier, I was able to fabricate a tool for doing this on a Saab-99 type front-wheel drive, using hard wood and following a picture of the tool included in an autobook abridgment. I was not able to fabricate the special screws needed to do this for the Austin America front-wheel drive. If you can't figure out a way to remove the clutch in such a car, don't do the job yourself (the dealer probably won't sell or lend you the special tool required). If the transaxle is in line with the engine rather than underneath it, as in a VW, there are no special problems.

Transmission Disassembly

With a transmission in front of you, the first thing to do (besides draining the fluid out if it hasn't been done), is to remove the top inspection plate or the side cover. If a side cover with shift forks is removed, put them both in neutral first. On a four-speed car with a top-cover gear selector do the same. As you remove the plate, observe the position of the forks and how they engage the synchronizer sleeves. Now look for fluid. If the transmission went dry and is burned and stripped, junk it and get a new one, a rebuilt one, or a used one to rebuild. The only thing I might reuse on a transmission that ran dry is the case.

Next turn the input shaft (clutch gear) and see if you can find the damage. Move the shaft up and down and notice the degree of movement in the front bearing in the case. If it moves noticeably in the case, the new bearing should be glued in with red Locktite. Be sure to make any end-play measurements recommended. Gears which stay engaged, like the clutch gear with the counter gear (layshaft), should not have rounded, worn, or chipped teeth. Gears that engage and disengage will have chamfered edges to ease engagement. The straight-teeth gears found sometimes on first and reverse gears are naturally noisy but stronger than spiral-cut gears. If you find a gear with chipped or worn teeth, also plan to replace the gear it engages because the damage will have been transferred.

The input shaft is supported by the pilot bearing or bush in the end of the crankshaft and the front bearing in the transmission housing. The main shaft is supported by either one or two sets of roller bearings or a bushing that fits inside the clutch gear. The nose of the main shaft goes right into the rear of the input shaft (clutch gear). The main shaft is also supported by one or two roller bearings in the rear of the transmission case and in the tail of the transmission. Most counter gears (laygears) are supported on the countershaft by needle bearings. Most reverse-idler gears are supported by brass bushes. Plan to replace all bearings as part of the rebuild.

Remove the speedometer gear from the tail piece and check it out. It is usually a soft gear, so that it will strip earlier than the gear on the main shaft. Each transmission has its own step-by-step disassembly procedure, as described in *Chilton's* or in a shop-manual abridgment. Petersen's book *Basic Clutches and Transmissions*, has good photographs of several common American transmissions being dismantled.

If the shifting forks are mounted on a plate that is removed, things are simple. The forks may be side mounted, but not on a removable plate. If this is so, then the pin will have to be driven out to allow the shift forks to be pulled away from their engagement with the synchronizer assemblies. If there are shift rods mounted in the case as in many foreign transmissions, then the shift forks have to be freed from the shaft so that it may be extracted. This requires loosening a set screw. I recommend Locktite on these screws upon reassembly.

Before extracting the shift rods be sure you have a blowapart diagram that shows the position of all detent springs, shift-rod balls, and interlock balls. If you don't have such a diagram, look over the case of the transmission for hex-headed plugs. Remove these, noting by some code or drawing where they go. Note particularly the interlock balls which go between the rods and make it impossible to move more than one rod at a time. Remove one of the shift rods, then the interlock ball, then the next shift rod. Then take the shift fork out of the case. At this point I recommend looking carefully at a blowapart diagram to see if there are any spring-loaded ball-bearing units such as the synchronizer units in a B.M.C. (British Motor Corporation) transmission or the free-wheel unit in the Saab 99. If these units come apart during disassembly, don't try to put them together again, but get a new unit at the dealer.

Inch-size ball bearings are available at a well-stocked hardware store, but metric sizes are unavailable, so don't lose those foreign ball bearings.

Input-Shaft Removal

To remove the input shaft, a bearing retainer has to be removed. The gasket for this unit should be replaced. Sometimes the input shaft pulls straight out, on other designs the tail piece and main shaft have to be pulled back so that the input shaft can be pushed into the case and then extracted through the side of the case. The bearing is retained on the shaft either by a large nut or a snap ring. The bearing is retained in the case by a snap ring. A mess of roller bearings will fall out when the input shaft and main shaft are separated. Collect them and count them. Examine the inside of the clutch gear where the nose of the main shaft fits. This is the bearing surface for the roller bearings. If it shows signs of wear it should be replaced. If the shaft is bent, replace it.

Cluster-Gear and Reverse-Gear Removal

The shafts that hold the cluster gear and reverse gear may be held in place by a pin or occasionally by a C-clip. Drive out the pin with a punch. Keep in mind that the shaft has to be put back, so the hole in the shaft lines up with the holes in the case. The counter gear shaft is usually driven out of the case with a large drift punch. Most counter shafts have a front and a back. Notice any grooves or holes that show which is which. You might leave the reverse gear in place if it shows no sign of damage. It doesn't get much wear.

The case and all other parts should be throughly cleaned. As you dismantle the main shaft, mark the elements as they come off. There is frequently a front and a back to the various gears which is not readily apparent in the blowapart diagrams. I put masking tape on each unit as it comes off giving it a number, and an arrow to designate the front.

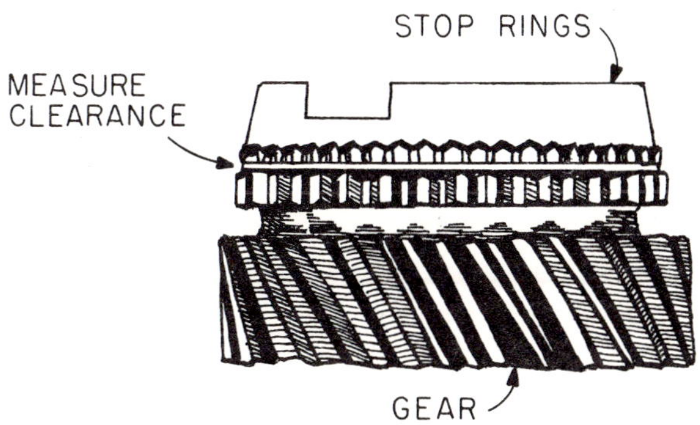

Synchromesh

The function of the synchromesh is to assist two gears into engagement that are moving at different speeds. Most transmissions in modern cars are supposed to have a synchromesh unit between all forward gears. Whether the one between first and second actually works, or to what extent, varies from car to car. Few VWs on which I have worked have had a functional synchromesh between second and first, enough though that I know it's supposed to be there.

With the exception of older Cadillacs, some Chrysler products, Porsche, and some BMWs, all synchromesh designs are similar. A synchronizing gear sleeve slides over a synchronizing hub pushed by the shift fork. The brass stop ring locks with the unit and is therefore turning as its speed. The inside of the stop ring is tapered and grooved; it fits over the cone of the speed gear (first, second, third, or fourth). As the stop ring engages the speed-gear cone, the stop ring turns the gear by friction bringing the gear into the same speed as the synchromesh unit, which is splined onto the main shaft. As the speed gear takes the speed of the synchronizer unit, the gear sleeve slides over the dog teeth of the speed gear, and the two shafts are effectively locked together one turning the other.

With the synchronizer unit off the shaft, the stop ring can be examined. Place it on the cone of its speed gear. There should be from .020" to .040" clearance between the edge of the stop ring and the shoulder of the section that has the dog teeth. The inside taper of the stop ring should be grooved. The tapered cone on the spring-loaded synchronizing unit, for all the stop ring will not engage on it smoothly. If the damage is not great, the edge of the tapered section can be ground off in a chamfer. Ordinarily it's a good idea just to replace the whole synchronizer unit. In many English cars the spring-loaded syncronizing unit, for all practical purposes, will have to be replaced as a unit because if it is taken apart, you—at least I—can't get it back together again.

The other thing to measure in the transmission is the end play of the gears on their shafts. This play is the front-to-back movement of the gear along the shaft. If this movement is too great, lateral pressure on the gear teeth will be too great. The ideal would be to have the gear shafts perfectly parallel with no front-to-back movement possible except in the sliding gears. This would give longer gear life. The acceptable end-play clearances are listed in a shop manual.

When rebuilding American transmissions you can purchase what is known as a "small-parts kit." This includes all the springs, roller bearings, and washers that should be replaced. You also want to replace the gaskets and probably the roller bearings and stop (synchronizer) rings. For foreign cars each part must be purchased separately. Go to the dealer (the only source for most of these things), have the parts man get out the blowapart diagram of your transmission with the parts numbers, and order everything you need from the picture.

Bring the car chassis number with you to aid in the identification. Bring damaged gear clusters with you to check teeth numbers with the new gears.

When the speedometer gear has to be replaced, measure its position so that it may be put back on in the same place. The speedometer gear has to be taken off to remove the rear bearing. A pipe, longer than the main shaft and with large enough inside diameter to fit over it, may be used to drive the speedometer gear off the shaft. Or a machinist can do it for you while you are getting parts.

Reassembly

The main trick of assembly is remembering how the unit was taken apart. There are additional tricks. When loose roller bearings are used on the counter-gear assembly, put the counter shaft into the gear, then install the individual bearings, each dipped in grease one at a time, around the shaft at each end of the gear. There should be enough bearings so that no holes are left. The shaft can then be removed. The grease will hold the bearings in place.

The thrust washers may be stuck on the case with grease, with the tang of the washer in the slot on the case. With the cluster gear roughly in place, start to ease in the counter shaft. It probably has a front and a back. Perhaps a notch or a hole is the indication.

Installing a main shaft into an input shaft that has loose roller bearings can be the most difficult step of the reassembly process. The loose bearings can be put into the clutch gear and stuck in place with grease. If the input shaft and main shaft are joined outside the transmission case, the problem is not great. But if you have to do this inside the case, try it outside a few trial times to see how it's supposed to fit. Don't force anything. With the parts together, rotate the input shaft and slide the synchronizer sleeves to see if they work.

When you put in the bolts that hold the tail piece, use a sealer on the threads as the holes in the case are tapped all the way through. If you don't do this, fluid will run down the threads and leak out. Do the same with any bolts holding a side cover in place.

Half-fill the transmission with fluid, then reinstall it in the car. If it is completely filled, fluid will probably leak out the tail during installation.

Used transmissions in my area are nearly unavailable from the junkies because a few private operators buy them all up. I've had some of my most interesting moments bartering and haggling with one of these transmission people. Sometimes I prefer to get one of their transmissions rather than rebuild the one I took out of the car; for example, if there is too much damage in the old transmission.

Transaxles

When you dismantle transaxles, you sometimes have to remove the differential gears. In these situations pay attention to any spacers on either side of the pinion-geared shaft, which engages the ring gear. If there is no damage on the ring gear and pinion, replace these units with the same spacers in the same places. The tapered bearings that support the pinion shaft, which is sometimes also the main shaft, may have to be tightened to a certain preload on the bearings. I put Locktite on the retainer nut. Preload is a measure of the drag, in pounds, on the shaft to rotation. On older VW and Austin America transaxles the preload is determined by measuring a gap with a feeler gauge. The Saab transaxle and the newer VW transaxles, which use taper roller bearings on the pinion shaft, have to have their preload measured with a special tool. I wouldn't try any shortcuts in the repair of the transaxle because it takes so long to get into it. The preload for new bearings is different from that recommended for old bearings. The dealer shop should be able to set this preload for you. He may charge you an hour's labor to do it, but if the total bill is going to be up to a grand, as it may well be for an exotic transaxle repair, then it's worth it. In the case of the VW, I would have the dealer, or a specialist, service and put together the whole pinion-shaft assembly.

You should be able to find an outfit in any larger metropolitan area which specializes in transaxle rebuilding. Until you have done a few regular transmissions successfully, it would be a good idea to take advantage of these services.

One of the special problems with front-wheel-drive cars is the universal joints that join the front axles to the differential. The Austin Ameria used a type which was easy to separate and replace; this was fortunate because replacement was needed often.

The older VW rear-axle arrangement was quite reliable. You may have to replace the boot covering the connection between the axle tube and the transaxle case. A split boot is available, so you don't have to dismantle the axle to get the new one on. The newer exposed rear-axle arrangement with constant-velocity joints at each end of the axle, does give troubles at times. A rumbling noise from the rear, especially on cornering, which may be heard with the engine off and the transmission in neutral is probably caused by the constant-velocity joints. The axle can be removed by taking out the six allen-headed screws at each end of the axle. With the boot pushed back and a C-clip taken off the end of the axle, the constant-velocity joint can be pressed off. Don't mix parts of this unit. If there is any sign of wear, get a new unit.

The Saab front axle is the most difficult to separate from its inner universal-joint connection. Some of the front suspension has to be dismantled. It is actually easier to separate the inner universals during engine removal. If you aren't removing the engine, then a tie-rod end has to come apart (see *Suspension and Steering*). For further information consult a shop-manual abridgment. If an engine mount comes loose with this type of drive system, the excessive torquing motion of the engine will destroy the right universal joint.

Differential

The differential gets its name from its function, which is to allow the different wheel-turning rates necessary on cornering. The inside tire turns less than the outside tire. If the car only went straight, a differential would not be necessary. Simple wagons with a solid rear axle manage to ignore this problem and simply allow the wheels to slip on the road surface. If you want to exercise your spatial-visualization abilities, look at a blowapart diagram of the differential and see if you can visualize how it works.

The ring gear and pinion gear, which are carefully matched, are the weak points of the differential. Actually, this part of the car is usually quite trouble free. Rarely a tooth will be knocked out of the ring gear giving a tick with each revolution of the tire. More commonly, the engagement of the pinion gear with the ring gear will be off, producing a howling sound or too much lash between the gear teeth. The howling may occur on acceleration (heavy heel contract) or on deceleration (heavy toe contact) or when the engine is floating at a constant speed (heavy face contact). Bad bearings will give a rough sound and enhance the noise mentioned above. With the rear end off the ground and the hand brake off, turn the tire back and forth. There is a little play in every driveline before a greater resistance to turning is felt. But if the tire can be turned 40 to 50 degrees with a "clunk" as the teeth hit, there is too much play. Check to see if the universal joints are in order. They would be a much likelier source for the problem. Adjustments in the ring and pinion will not get rid of noise caused by wear. The ring and pinion have to be replaced.

The ring and pinion gears are run together on a special machine at the factory. The best setting is determined by experiment. They are then marked to indicate how they vary from standard and shimmed accordingly. Adjustments are made in two directions. The pinion shaft can be shimmed to move more in mesh with a ring gear, or less, and the ring gear can be shimmed to move left or right. Older *Chilton's* editions set the ring and pinion gear by using red lead paint. Methods using a dial indicator are described in many shop-manual abridgments.

Adjustment being quite difficult, and the unit being fairly reliable and therefore cheap and plentiful on the used market, I recommend just getting another whole rear end at the junk yard. The main problem here is to find an exact replacement, or at least a replacement that is useful.

Since the ring gear and pinion gear have a reduction ratio, in the same way the transmission gears have reduction ratios, it is possible to get a ratio in the rear end inappropriate for the transmission or the engine capacity. The greater the rear-end ratio, the greater the leverage and the greater the rate of acceleration possible, but the usable top speed will be lower. A smaller ratio will produce a lower rate of acceleration, but a higher cruising speed. The smaller ratio will give better gas mileage generally, but you may feel that you are starting in second gear.

A rear-end ratio is determined as follows: Lift the rear of the car off the ground. Release the hand brake and put the transmission in neutral. Mark the tire with a chalk. Turn the wheel opposite some convenient starting place, like opposite the edge of a fender. Then have somebody count the number of revolutions made by the drive shaft as the wheel makes one full revolution. It will be somewhere between 3 and 4.5 revolutions. This is the ratio of the rear end.

Special rear ends may require a special high-pressure lubricant. Be sure to check on this. Otherwise 90-weight oil is used in the differential. Overfilling may cause leaking at the rear-axle seals. There is a seal in the nose of the differential, which can leak. Often this seal can be replaced by separating the driveshaft and removing the pinion-shaft flange. The old seal can then be pried out with a screwdriver.

Rear Axles

Semifloating rear axles are common on automobiles. In this case the axle is mounted in bearings inside the axle housing, with the wheel mounted on a tapered shaft or a flange. The axle is held in place by a bearing plate that also bolts through the brake backing plate through a flange on the end of the axle housing. If nothing appears to be holding the bearing in place, then in some cases the axle is held by a C-clip at the differential end of the axle. This design requires removal of the differential cover plate.

Some axles are a tight fit at the differential end and make it necessary that they be banged out with a side hammer attached to the flange end. A pair of claw hammers prying at the back of the flange and opposite each other will probably also do the trick. Mark the left axle and the right axle because their lengths are probably different. A press-fit bearing on a shaft can often be removed by grabbing the end opposite the bearing with a rag, and giving the shaft a wallop, hitting the bearing end against a wooden block on the ground. Otherwise a machinist can remove it quickly. If you have to tap the threaded end of the shaft with a hammer for any reason, put the nut on nearly all the way first to protect the threads.

The weakness of the semifloating rear axle is that it has to take more than just the driving torque. This axle also has to take bending caused by carrying the load, hitting ruts, traction, and breaking. If it breaks, you've lost your wheel. In the fully floating rear axle, the axle only takes the driving and breaking torque. The hub is mounted on bearings that are mounted on the axle housing. The axle housing therefore takes all the loads except rotational torque. However, a fully floating rear axle is only necessary when the vehicle is used under very rough conditions. Normal repairs to brakes and so on take twice as long on the fully floating axle rear end than on the semifloating rear axle.

With independent rear suspension, the axle is usually exposed. It will be articulated at least at the differential housing and probably at both ends. The articulation may either be universal joints or constant-velocity joints.

Chapter 17:

Suspension, Steering, Exhaust

Ye who, borne about in chariots and sedans
Know no fatigue but that of idleness.

—William Cowper (1731-1800)

The discussion in this chapter concerns subjects that I didn't feel deserved a chapter of their own, considering my more limited knowledge of them: the suspension system, the exhaust system, the body, automatic transmissions, and some other subjects.

Suspension System

The shock absorbers of the suspension system have a comparatively short lifetime—from 20,000 to 50,000 miles—but they are frequently ignored and left on for the life of the car. The lifetime of the rest of the suspension system is shortened if bad shocks are used for long. I am speaking here of airplane-type shocks, not the type used on General Motors cars in the 1940's and English cars of a more recent date. Those knuckle-action shocks were supposed to be checked, and filled with hydraulic jack oil when low. If they weren't allowed to run dry, they lasted as long as the rest of the suspension system.

Although I have tried various tricks to test the air-plane type shocks while they were still installed, I have found that for a sure test they have to be removed. Take the shock in hand and pull it all the way open, then all the way closed. It should move slowly and smoothly at the same rate in both directions, without roughness and without squeaking or gurgling noises. The steering-linkage damper on the steering mechanisms of some cars wears out in only about 1" of its normal travel, but it is there that it spends 90% of its time.

The primary function of the shock absorber is not to absorb the shock of hitting something with the wheel. It is rather to keep the spring, which does absorb the shock, from continuing to bounce up and down after the initial blow. The shock absorber damps the oscillation of the springs. It does, however, help a little to soften the initial shock, and also it helps in cornering. There is no particular problem in removing shock absorbers unless they are rusted on (call for a torch) or mounted inside of a coil spring with no way to unbolt them from the bottom. You have to keep the top of the shaft from turning while you loosen the lock nut.

If the shock is mounted inside a coil spring or if you are dealing with MacPhearson struts, you may have second thoughts about replacing them yourself. A coil spring is like a loaded bomb. You wouldn't want to be in the way of a coil spring, when it lets go. K-D makes a coil-spring squeezer, which enables you to compress the spring and remove and renew the shock absorber. If the suspension drops down far enough—as is usual with the rear end—to take all the tension off the spring, then no special tools are ordinarily needed to remove it. Most shocks mounted within coil springs can be removed through the bottom of the coil without trouble.

Another type of spring is the leaf spring, the original buggy suspension. It is composed of a stack of spring steel leaves attached at the two ends by shackles to the body. In the middle they are bounded by U-clamps to an axle housing. Leaf springs should have a curve in them. If they are flat, they are finished and will probably break with the next bad pothole. Leaf springs of just about any kind can be rebuilt by outfits that specialize in these things. Check the Yellow Pages under "springs". You can have them built to your order and have extra leaves added if you like.

Torsion-bar suspension is also used and is becoming more common. It seems to be reliable. The torsion bar is connected in such a way that upward movement on the wheel twists the torsion bar, which opposes this twisting movement. VW uses torsion bars for both front and rear suspension. Torsion bars are also used to oppose the twisting movement of the differential housing upon acceleration. Such a bar would be connected to the axle housing and the frame. The rear torsion suspension on the VW can go out of adjustment, especially on the bus. If the rear of the bus is low over the rear wheels with no load, then the torsion arms are in need of adjustment. Refer to a shop manual abridgment for details.

There are other parts of the suspension, such as various struts and stabilizer bars, which help control the movement of the wheels. Rubber bushings are used with these and may need replacement after ten years or more.

Engine Mounts

Another type of suspension, or should I say cushioning, are the rubber mounts for engine and transmission. The function of these mounts is to cushion the vibration created by the engine. The mounts electrically isolate the engine from the body, hence a ground strap must be connected between the two.

The mounts have a tendency to be overlooked, but if the clutch-pedal linkage collapses on acceleration (the type with a shaft between the body and the engine block), or if you are destroying right-side U-joints on a front-wheel-drive car, or if the engine sits strangely, you probably have defective engine mounts. To check, watch the mounts while somebody gooses the accelerator to torque the engine. If there is undue movement, replace the engine mounts. To check the alignment of the engine, place a carpenter's level across the mouth of the carburetor. It should be roughly parallel with the ground.

Front-End Analysis

In areas of the country where the roads are good, front-end alignment or realignment is infrequently needed. In other places or on vehicles driven off the road it is more common. It is difficult to predict the case of a particular alignment problem, because any of the elements involved in front-end alignment can produce any of the symptoms.

Some of the symptoms of front-end trouble are shimmy or looseness transmitted through the steering wheel, "tramping" (wheels walking down the road after hitting an obstacle rather than rolling), and "wander", (difficulty in keeping the car rolling straight). Sometimes after hitting a road "imperfection," especially at freeway speeds, a resonant vibration in the front end can be created. Give the brakes a sharp blow (not enough to stop the car appreciably), and this will probably stop the vibration. It is important to get any reasonant vibration damped anywhere in the car, whether in the crank shaft or the front end—nothing will destroy machinery faster.

The owner's report of the symptoms may be very imprecise, maybe totally inaccurate. Considering these difficulties it is important to have a reasonable diagnostic procedure.

The first step is to be able to reproduce the symptoms. If you can't reproduce the symptoms, don't try to fix the trouble, because you won't know in the end if you were successful. As you are driving straight ahead, notice if the spokes of the steering wheel are symmetrical. If not, it could be that the steering wheel was removed, then put back in the wrong place (an annoying problem but not serious) or it could be that the steering linkage needs adjustment.

Find a railroad crossing to drive over at 30 mph or so. This should give you the best instance of a front-end problem. Of course, different cars ride differently. The big-size American car generally floats along very comfortably, but normally doesn't corner well, whereas some of the small B.M.C. vehicles ride like trucks normally, but corner better.

After a road test, check the tires. Improper in-

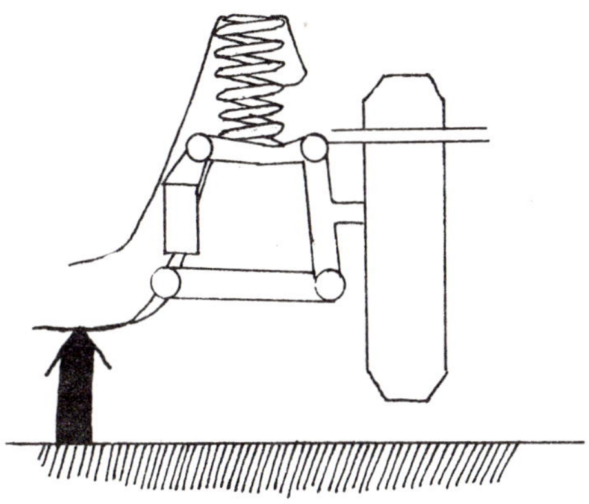

flation or wheel balance can be the source of the problem. If steering-wheel shimmy is the problem, and the car normally has a steering-linkage damper (VW), then replace the damper. Let somebody roll the car down the road straight ahead and walk behind it noticing if the rear wheels are following in the same track as the front wheels. Look for any wheel wobble.

If the tires are not tracking, pick up the rear end and inspect the mounting of the rear axle with the springs. It should be at right angles to the drive shaft. The frame may be bent and cause tracking problems. This can be checked by marking a point on the road with chalk just under the right front frame and the comparable point on the left side. Then make spots on the road under two comparable places of the left and right ends of the rear axle. Move the car. Now measure the distances between the left front mark and the left rear mark, the right front mark and the right rear mark. If they are different by more than 1/4", then either the rear axle is out of alignment or the front suspension is pushed back on the side with the shorter dimension.

Now measure the distance from the right front spot to the left rear and from the left front to the right rear. If these diagonal measurements differ, the frame of the car is diamond shaped, that is to say, bent. You have to go to a frame-straightening specialist or even consider junking the car if it isn't worth much.

If the car tracks alright with no wheel wobble, check and replace the shocks if necessary (see previous section). If the wheels wobble, pick up the front end and inspect the joints and links comprising this unit.

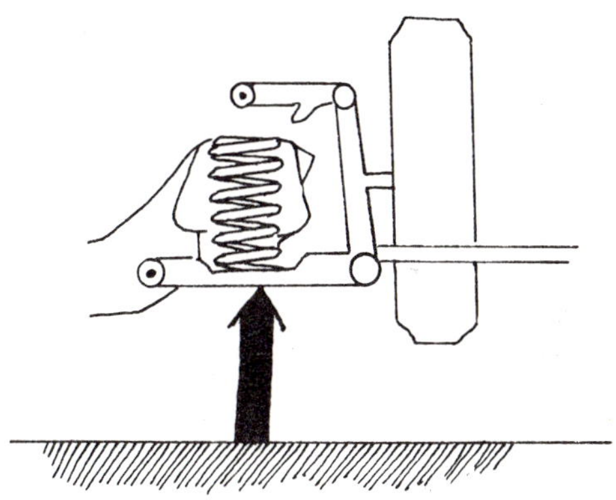

For this inspection, the lifting should be done in either one of two ways, so that the pressure on the load-carrying ball joint is relieved. If it is not, it will seem tighter than it is. Grab the wheel top and bottom, and rock it back and forth and lift it up and down. If there is no play, proceed to the next step. If there is play, check first the axle bearing (see section on wheel bearings in *Brake* chapter). If this is in order, then have someone wiggle the wheel for you and observe the king pin (on an older car) or the ball joints. Little play is allowable, but it seems commonplace to have ball joints replaced when it is not necessary. So check on the allowed play in a manual for the car or in *Chilton's*. A dial indicator is a useful device for measuring play.

Ball joints can be a real pain to replace. Some have to be pressed out of the control arm. At least a "pickle fork" (tie rod-end remover) will be necessary.

Next, grab the two front tires by the front and move them in and out together. Observe the movement in the drag link, the idler arm, and the tie rods. You should see no movement in the joints. New tie-rod ends, idler arms, and other bushings can be purchased at your parts store.

If the front-end problem persists after you have eliminated the tires, tracking, wheel bearings, shocks, and steering connections, as causes, it is time to do a wheel alignment. I send wheel-alignment problems to a specialist, but I will review the terminology of wheel alignment here.

Wheel alignment is the position of the wheel in relation to a flat surface: the road. At first we would probably think that the tire would be best oriented straight up and down, and the two front tires parallel. It turns out that to compensate for various forces experienced during driving it is best to cant the wheels ever so slightly off center. Following are some terms used in alignment:

"Caster" is a slight inclination of the king pin (an imaginary line drawn through the centers of the ball joints) forward, or back in some cases, to help stabilize the steering. Unbalanced caster can cause the car to pull to one side or the other, or cause hard steering.

"Camber" is a slightly outward tilt to the top of the tire as seen from straight ahead. Its purpose is to put more of the car's weight on the inner wheel bearings and to compensate for looseness and wear in the steering knuckle and wheel bearings.

On older cars caster and camber were often adjusted by an eccentric upper-control-arm pivot, but on newer cars they are often adjusted by shims spacing out the upper control arm.

"Toe-out" refers to a difference in angle between the two front wheels when they are turn-than in the rear; just like a pigeon-toed person. Toe-in compensates for the tendency of the wheels to splay out in front; it is adjusted by changing the lengths of the tie rods.

"Toe-in" refers to a difference in angle between the two front wheels when they are turning. The wheel on the inside of the turn makes a slightly greater angle than the wheel on the outside. This is determined by the shape of the steering arm to which the tie rod connects, and is not adjustable, only replaceable.

Steering Box

The steering box is usually trouble-free. But when it runs out of lubrication it will 10,000 miles later grind itself to death or the bolts that hold it to the frame will sometimes loosen. If steering becomes extremely difficult with funny "spronging" (spring-exploding) noises, the steering box may be loose. There are three steering-box adjustments, but these should not be attempted until a thorough inspection is made of the steering linkage for play, with worn parts replaced first.

Often the steering shaft couples with a short input shaft going into the steering box. The shafts should be parallel and in line. If not, jack up the front of the car and loosen the bolts holding the steering box to the frame to allow the steering box to align itself. You may have to shim the steering box, if there is a gap at any of the mounting bolts.

The next adjustment is necessary, if there is end play in the "worm." The worm is part of the input shaft and has a bearing on the end of it whose position is adjusted by shims or an adjusting nut. If there is any play in the steering wheel when you pull it out toward you and away from you, this adjustment should be made. While making this and the following adjustment, disconnect the drag link from the pitman arm. A pickle fork will probably be necessary for this. Back off the roller-shaft adjusting screw. The end-play can be adjusted in some cases by an adjusting screw held in place by a lock nut, or by removing a plate and changing the number of shims abutting the bearing. The adjustment is proper when there is just the slightest resistance to turning in the steering wheel (1 lb.).

The final adjustment is in the lash between the worm and the cross shaft. The lash is felt as play in the steering wheel when everything is hooked up. There should be no lash, or at least very little. To adjust the lash, free the lock nut of the cross-shaft adjusting nut. Turn the steering wheel from full right turn to full left turn, counting the turns. Return to the half-way position by dividing the number in half. In this middle position, screw in the adjusting screw until there is the slightest increase in drag. Turn the wheel back and forth through the middle position. There should be a slight increase in drag as you pass through the middle position. Tighten the lock nut and reconnect the drag link to the pitman arm. Check the steering box for fluid.

If you feel roughness when turning the car, or a "dead spot" in some position of the steering wheel (caused by a bad roller bearing), or if you have rack-and-pinion steering, then the steering box has to be removed. This requires (1) freeing all steering-column supports and the input-shaft to steering-shaft coupling so that the steering column can be pulled back, (2) disconnecting the drag link, and (3) unbolting the steering box from the frame. If the steering arrangement has an unbroken shaft extending from the steering wheel to and including the worm gear, it is necessary to have a lift to get the car or truck at least four feet off the ground. I pass these jobs on to someone else. On the other hand, I drove a Ford truck for four years with a dead spot in the steering and it never caused trouble or got worse.

In my experience it is best to get a steering box from a junked car for replacement. Check to see whether it still has fluid and whether you can turn things without embarrassing noises coming from the box. Rack-and-pinion adjustments are made by changing the shims under the bearing preload cover and under the yoke cover. Check a manual for details.

Exhaust System

I send my customers who need tail-pipe, muffler, or exhaust-pipe repair to a small muffler shop, which does a better job for less than the national giant in muffler repair. The big-chain muffler shops underpay their help and make a bundle on the parts. Underpaid help is poor help.

For practical purposes, exhaust problems start with the exhaust valve (see *Valve*). Then comes the exhaust manifold with its gasket. Many American cars are designed to do without an exhaust manifold gasket; but if there is a slight warp in the exhaust manifold, you can buy a gasket from a parts house made for Ford cars, which you can't get from the dealer. The gasket may solve the problem of a recurring leak. If it doesn't, you can sometimes get your machinist to resurface the manifold. If the flange gasket (where the exhaust pipe connects to the exhaust manifold) keeps blowing out, the flange could be warped. The flange can also be resurfaced when the studs, if any, are removed.

If the exhaust manifold is cracked, there is no cure except welding. The welding has to be done by an expert because the manifold is cast iron. The welding may cost as much as a new manifold. Check also the junkyard.

There are usually special nuts used on the exhaust manifold. They are either brass or they have a special thread insert to prevent them from galling onto the bolts or studs. Keep these nuts separate when you do a general disassembly, so they will be used on the exhaust system when things are put back together again. Corrosion can reduce the outside size of the exhaust-manifold nuts. I remember a Falcon where the nut size was nominally 9/16" American standard, but I also had to use 14 mm and 1/4" Whitworth sockets to get a tight fit on nuts reduced in size by corrosion.

The exhaust manifold often has a heat-riser valve. This valve directs the flow of hot exhaust fumes either close to where the carburetor is mounted (the hot spot) or farther away. The path of the exhaust is controlled by a flap. This is opened and closed by a bimetallic spring and a counterweight outside the manifold mounted on the flap shaft. This device creates a hot spot under the carburetor to assist in vaporization of the gas/air mixture while the engine is cold. After it heats up, the hot spot should be bypassed or the gas mileage may go down, as well as the performance.

Testing this unit consists in rotating the counterweight to determine if there is free movement. WD-40 would help free this unit. There is a foamy aerosol spray made especially for this part, which is available at Chrysler dealers. The dealer will also be able to give you new parts for this unit.

Exhaust Fittings

Another possible cause of continued flange-gasket failure, besides a warped exhaust-manifold flange, is a loose exhaust pipe. If the hangers are broken, the pipe can rattle and greatly accelerate pipe breakage, joint separation, and gasket destruction. For this reason check the exhaust pipe and tailpipe hangers.

Low spots in the system, such as a crossover pipe on a V-8, will rust rapidly. Water, being one of the products of combustion, settles in a low spot. Cars that have been sitting will have their exhaust system rust in a hurry because the metal is unprotected by paint and not being dried out by the hot gases of the running engine.

Automatic Transmission

The automatic transmission is another device that I leave to a specialist for repair. It does not present greater difficulty than, say, engine rebuilding, but I have had fewer occasions to work on it. An unusually high number of my customers have standard transmissions. It seems to me that a detailed shop manual would be useful if you wish to get into this part of the car. Petersen's *Basic Clutches and Transmissions* has a good introduction to automatic-transmission overhauling.

There are some aspects in automatic-transmission repair that are part of general knowledge on car repair.

When you check the transmission-fluid level, the engine should be running and warm. It is important to have the correct fluid level. If the fluid appears brown and discolored, as opposed to red, then it should be drained and replaced. Fords made in 1967 or earlier should use fluid marked F-M2C33-D. Later Fords use F-M2C33-F. Most other cars use Dexron Type A for cars produced in 1967 or before, and Type B on later cars.

Automatics should be periodically drained and cleaned. The bottom pan has to be dropped off. Since there is usually no drain plug, you have to remove the pan with the unit still full of fluid. If you can figure out how to do it without making a mess you are better than I. After the pan is removed, a filter will be exposed. Some automatics have replaceable screens, others use a screen that you remove and clean. Thoroughly clean the pan and screen, if it is the nonreplaceable type, and scrape off the old gasket. Provide a new gasket, a new filter, and clean new fluid and you've completed the maintenance routine.

There are often three connections to the automatic transmission, which may be the cause of poor shifting. The linkage from the shift lever on the steering column may be out of adjustment. This may be checked by putting the shift lever in neutral, then going under the vehicle and disconnecting the shift rod from the shift arm on the side of the transmission. Operate the shift arm by hand, putting it in its neutral position. The shift rod should now line up with the shift arm. Change the length of the arm if it does not.

Also, there may be mounted on the carburetor or throttle linkage a push-button on-off electrical switch, which operates a downshift valve. It may also be operated by a mechanical link from the throttle to the transmission. This valve automatically shifts down a gear, no matter which one you are in.

Finally, there is a throttle valve or, on later models, a modulator valve. The hydraulic pressure in the clutches, bands, and valves is regulated in part by this valve. If there is a harshness or slippage with shifting, then this valve may be faulty. On older cars this valve was controlled by a mechanical link, the throttle-valve rod, from the accelerator linkage to the transmission. Consult *Chilton's* for accurate adjustment of this rod. Nowadays, the valve is controlled by vacuum. A vacuum hose runs from the intake manifold to the modulator valve, which is screwed into the outside of the transmission. Check the hose for leaks and replace the modulator valve if you have shift problems.

If the automatic transmission has to be removed, you need a transmission jack because this unit is too heavy for removal without it. After an access cover is removed from the bottom of the transmission, bolts can be removed that hold the torque convertor to the flywheel. Wedge the ring gear with a screwdriver while loosening these bolts.

Now that my mind has been wrung dry of all information it contained on repairing automobiles, I would like to indulge myself with a short story inspired from my experience as a mechanic. It concerns my approach to building an ark.

When you build an ark, it is important that it won't sink. This means devising a construction which won't leak. And so, armed with good intentions, the correct tools, and quality materials, we begin the work.

Then around half-way through we begin to make compromises. Sometimes we say to ourselves, "That'll work out OK" without realizing what we have done. But the fact is, reaching perfection is costly—too costly. So the best we can hope for is that the ark outlive our need for it.

Nevertheless, a small voice in the background asks: "What about the small leak?" to which the answer comes: "That's what the sponge is for, mate."

With that, I throw in the sponge.

Mechanic's Tips

Pinch clamps, valve lock nuts, screws going into aluminum are commonly overtightened. Gasline and brake-line leaks will never be stopped by overtightening. If you are worried about a screw or nut getting loose, use Locktite.

Take the old part with you to the parts store and compare it with the new part. In electrical parts with several wires, cut the wires as they come out of the damaged part, leaving the old wires hooked up so that the new wires can be replaced easily one at a time.

Some screw positions are best remembered by putting the screws right back in their holes after a unit has been removed—others may be safely gathered in groups in tin cans.

Automatically replace short pieces of hose in hard-to-reach places, such as between the head and block.

Replace gaskets wherever possible on reassembly or use silicon sealer. Use sealer on bolts that are to go into holes taped through a housing to a place where oil or water goes, such as the bottom bolts of a transmission housing.

Always replace cotter pins.

Use Locktite instead of reusing locktab washers.

Copper washers seal, so save all washers of any kind left over from kits. The copper washers, for example, from the VW oil-strainer kit can be used to shim the needle valve in a SU carburetor.

Don't forget to tighten the wheel lug nuts after lowering the car to the ground.

Put oil in a newly rebuilt engine.

If the car doesn't start after you replaced the points in the distributor, you probably forgot to put the rotor in.

If the car doesn't start, look to the last place you worked on the car first.

Replace two sliding parts as a pair if possible.

Replace both members of a pair of mating gears even if only one appears damaged.

If you put a ding in the oil pan, it will warp, making it difficult to put on and assuring a leak to follow.

Two gears, directly meshed end to end, turn in opposite directions.

Two gears, turned by a chain, turn in the same direction.

Two gears, directly meshed, turning shafts at right angles to one another (e.g., the distributor shaft and camshaft), turn either in the same direction or the opposite, depending on which side the gear mesh on.

Learn the smells of gas, engine oil, differential oil, and brake fluid.

Whenever working with bolts or pulleys, check alignment.

Never bang on the end of a shaft that has a roller or tapered bearing mounted on it while it is mounted in its housing; the bearing will be damaged.

Disconnect the ground strap of a battery when working on the electrical system of newer cars (e.g., cars with solid-state electrial components).

Re-educate the driver after doing major work on the car.

Squeaking that develops after the car is running a while comes from a rubber part.

Squeaking that goes away after the car is running a while comes from a bearing going dry.

If there is a question of bearing or gear deterioration, take some of the lubricant and rub it into a cleaning rag and look for metal flakes on the surface of the rag. Metal flakes indicate bearing deterioration.

A spring, when mounted anywhere in the car and in its rest position, should be just barely extended. If it is loose or stretched tightly, the spring is mounted in the wrong hole or is the wrong spring for the application.

Use antiseize on the threads of plugs or bolts going into aluminum, especially if the aluminum has been cleaned and exposed to the weather.

Don't accept another person's diagnosis of a problem. Do your own diagnostic routine if just to experience the symptoms.

When you evaluate an older car, don't consider so much its name or the rumors about its supposed strengths or weaknesses, but how well it was taken care of and how many miles it has. But watch out for the "unique" car for which you can no longer get parts.

Every grade of petroleum except tar can be considered a solvent for something. Oil would be a better solvent for tar than something volatile. Gas could be used to clean up the oily residue left by kerosene.

When doing an engine job on a car with air-conditioning, expect to spend more time. The unit should be dismounted and moved aside rather than disconnected. Timing is often difficult with an air conditioner in the way.

Don't do $200 worth of work on a car worth $100.

Safety Rules

Before starting the engine, set the parking brake and place the gear selector in neutral (standard transmission) or in park (automatic transmission). Never stand in front of a car being started.

The carbon monoxide in exhaust gas is highly poisonous. To avoid asphyxiation, operate the vehicle in a well-ventilated area. If the vehicle is in an enclosed area, the exhaust should be routed directly to the outside by a leak-proof exhaust hose.

When using a timing light in a garage or other enclosed area, keep the light at least 18 inches above the floor to minimize the possibility of igniting gasoline vapors. When operating any test instrument from an auxiliary power source in a garage or other enclosed area, place the power source at least 18 inches above the floor to minimize the possibility of igniting gasoline vapors.

The gas formed above the electrolyte of a charging battery is explosive. Do not smoke or permit a flame or spark to occur near a battery on charge. Also, make sure that test equipment is turned off when making a connection to the battery.

Keep hands, hair, necktie, loose clothing, and test leads well away from fan blade, fan belt, power-steering belt, air-conditioner belt, and other moving engine parts; serious injury could result from entanglement.

Do not touch the hot exhaust manifold, the radiator, the spark plug, or coil terminals. Spark voltages are not normally lethal, but an involuntary jerk of the hands caused by electrical shock may result in injury.

To avoid the possibility of a flash fire, do not smoke or permit a flame or spark to occur near a carburetor, fuel line, fuel filter, fuel pump, or other potential sources of spilled gasoline or gasoline vapors. If a small fire starts in the carburetor, smother it with a rag.

Never look directly into the carburetor throat while the engine is cranking or running. A sudden backfire can cause serious burns.

An automobile battery can produce very high currents. Therefore, be careful when working near the battery to avoid electrical connections through tools, wristwatch, and so on.

Avoid contact with the battery acid. It can eat holes in clothing, burn skin, and cause permanent damage to eyes.

Never remove the radiator cap while the engine is hot. Hot coolant escaping under pressure can cause serious burns.

The jack supplied with the vehicle should be used only for changing wheels. Never crawl under a car or run the engine while the vehicle is on the jack. Use jack stands after the car is raised.

Have an eye-wash cup available for cleaning eyes, and a pair of tweezers for removing metal splinters.

Glossary of Technical Terms

allen head	A hexagonally shaped recession in a bolt head requiring an allen wrench to loosen or tighten.
arcing	A machining process whereby a brake shoe has its surface ground into a smooth arc.
A.T.D.C.	After Top Dead Center (see T.D.C.)
bearing	A part in which a journal, gudgeon, pivot, pin, or the like turns. Implied is the capacity to incline. This general term covers roller bearings, tapered bearings, bushes, and paired washers.
bezel	A sloping edge. Refers to the sloping edge or face of a cutting tool, and also to the grooved rim or flange in which a glass crystal is set, for example, a speedometer glass.
big end	The end of the crankshaft to which the flywheel is attached. More weight is carried, and the bearing is commonly larger than the opposite end.
bore	The hole or cylindrical cavity thus formed. The diameter of a bore is measured with an inside micrometer.
bottom end	The crankshaft plus the crankcase plus their attendant paraphernalia.
broach	In machining, a slightly tapered cutting tool used for accurate sizing of a hole. The material is first drilled to a few thousandths of an inch undersize, then broached with a broach or a reamer. This is done with valve guides and wrist-pin bushings in the connecting rod.
brush	One of two or more carbon pieces, which bear or brush against a commutator or slip ring, providing for passage of electrical current. They are found in any electrical motor or electrical generator.
B.T.D.C.	Before Top Dead Center (*see* T.D.C.)
bush	A metal lining that has special surface properties; namely those of a bearing, capable of enduring. Metals such as bronze silicone and Babbit or White Metal are slippery and are frequently used in bushings.

cam	A rotating projection used to impart movement in an exactly timed fashion to a roller, follower, or lifter.
camshaft	A shaft that has all cams needed to operate the valves and the fuel pump on the internal-combustion engine.
C-clip	A steel clip that fits into a recess cut in a shaft, used to hold in place things mounted on the shaft.
chamfer	As a noun, the surface formed by cutting away the angle formed by two faces of a piece of metal or the act of making such an angle; as a verb, to bevel.
compression ratio	The ratio between the free volume of the cylinder at Bottom Dead Center to the free volume at Top Dead Center.
cotter pin	A split metal pin, the ends of which are bent after insertion through a slot or hole.
crankshaft	A shaft with turns or bends.
cross thread	The threads of a bolt and the threads of a tapped drilling, cutting each other because they were started at the wrong angle. The old threads will be cut away and new threads cut in their place. These new threads are much weaker than the original.
detent	That which unlocks a movement; a pawl, a dog, a click.
detonation	The instantaneous combustion of the fuel mixture, as distinct from the progressive passage of the flame front through the unburned mixture in normal combustion.
diode	A small solid-state component, which allows the flow of current in one direction only.
dog teeth	On a gear, teeth that are chamfered along all their edges.
drillings	Bores or holes made by a drill. Refers to holes made in a casting, such as a carburetor.
dwell	The time the points are closed and the coil is charging, before one of the distributor camshaft nodes opens them. It is measured in degrees.
dynamo	The British term for electrical generator.
flange	A rib or collar, for strength or guiding, or for attachment of another part.

flare nut	The nut that fits round brake lines or gas lines that have been flared or opened up. The nut holds the flared line against a brass seat at the bottom of a threaded drilling.
fly cutting	A circular cutting process used to resurface a recessed circular area, such as a recessed flywheel or the heads on a VW engine.
galling	A wearing away of metal by friction, when lubrication breaks down. Some metals have a disaffinity for one another and will bind together. An attempt to separate them (unscrew) will result in breakage due to galling.
glass-bead blasting	A process of blasting metal parts with a high-pressure air stream containing tiny glass beads. This is a safe cleaning process for aluminum parts because it doesn't remove or pit the metal.
grinding	To reduce to powder by a grinding wheel. When the crankshaft or camshaft is reground, the wear surfaces are reduced in size to remove scores, grooves, or pitting. The new surface is then polished and hardened if necessary.
gudgeon	A pivot or journal.
hardening	A chemical process whereby the surface of a steel part is made harder and more enduring. This hardening affects a few thousandths of an inch of the surface. When this layer is worn off, metal deterioration accelerates.
hex-headed	The standard six-sided machine-screw head or nut.
hone	A tool for enlarging holes to precise tolerances with a mechanically rotated and expanding set of fine-honing stones.
journal	The part of a rotating axle, shaft, or spindle that turns in a bearing.
knock	*See* detonation.
land	In any surface prepared with indentations or grooves, the part *not* so treated, as the surfaces between the ring grooves on the piston.
lash	The gap between two engaged gears or the space between the valve end and a rocker arm; a lash results from an abrupt movement.
magnafluxing	A process that helps to make visible small cracks in metal parts.
manifold	A pipe fitting for connecting one pipe with several others.

pitman arm — Originally a rod that connected a steam engine to whatever device was controlled by the pitman. Now it refers to the arm that connects the steering box to the steering linkage.

port — An opening for the inlet or outlet of air, gas, or water.

preignition — The ignition of the fuel mixture by an abnormal means (such as a glowing speck of carbon) before the spark-plug ignition occurs.

reaming — *See* broaching.

relay — An electromagnetic switch that is operated by variation in the electrical circuit and that in turn operates other devices.

r.p.m. — Revolutions per minute.

runout — The degree to which something is not turning on its intended axis. A measure of wobble. It is measured with a dial indicator.

sand blasting — Similar to glass-bead blasting except that sand is used. This process is more abrasive.

seized — To take or lay hold suddenly or forcibly; to grasp. As when a bearing shell seizes a journal.

shim — As a noun, a thin piece of metal, perhaps in the shape of a washer, used to fill out a space. As a verb, to effect small adjustments in clearance.

short block — The lower portion of the engine; the crankcase, and most of its parts.

short to ground — An electrical short circuit that occurs when a hot wire from the battery breaks and touches an unpainted body part or the engine.

solenoid — An electromagnetic switch.

spider — The British term for U-joint.

splined — Grooves cut longitudinally in the end of a shaft, which mate with a hub.

stake — To take a hammer and a chisel and fix a nut on its bolt or metal cover in its housing by deforming the metal in one or two spots where the two parts meet.

surfacing — A machining process whereby a new level surface is ground on a part.

tach-dwell meter — A meter that may be used both as a tachometer and a dwell meter. The tach measures engine speed in r.p.m.; the dwell measures in degrees the period in which the points are closed previous to each spark-plug ignition.

tang	A projecting shank or tongue, as on a knife, file, or chisel on which the handle is placed.
tap	As a verb, cutting a female thread into a drilling by means of a light blow; as a noun, the tool to do it.
T.D.C.	Top Dead Center. When the piston reaches the very top of its upward movement in the cylinder.
torque	The action that produces or tends to produce rotary movement, measured in foot-pounds.
turning	A machining process in which a cutting blade is used to resurface the drum or the rotor of a brake system.
venturi	A short tube with a constricted portion. Pressure decreases at the constriction as velocity increases. This is true for air as well as water.
whitworth	A little-used sizing system of threading nuts and bolts of some early British cars. As opposed to metric or American Standard.
worm gear	A gear shaped like a spiral along the axis of the shaft.

Bibliography

Ball, Kenneth. *Autopress*. Brighton, England: Autopress Team of Technical Writers, n.d.

Basic Automotive Tools. Los Angeles: Peterson, 1974.

Basic Clutches and Transmissions, 3rd ed. Los Angeles: Peterson, 1974.

Blanchard, H.F. and R. Ritchen. *Auto Engines and Electrical Systems*, 6th ed. New York: Motor, 1973.

Campbell, Colin. *The Sports Car*. Cambridge, Mass.: Robert Bentley, 1955.

Chilton's Auto Repair Manual. Philadelphia: Chilton's Book Co., published annually.

Dyke, A.L. *Dyke's Automobile and Gasoline Engine Encyclopedia*, 8th ed. St. Louis: A.L. Dyke, 1918.

Motor's Auto Repair Manual. New York: Motor, published annually.

Motor Truck Repair Manual. 28th ed. New York: Motor, 1975.

Muir, John. *A Manual of Step by Step Procedures for the Compleat Idiot*. Santa Fe: John Muir Publications, n.d.

Nelson, Carl A. *Mechanical Trades: Pocket Manual*. Indianapolis: Theodore Audel, 1974.

Power and Hand Tool Catalog. Chicago: Sears Roebuck.

Volkswagen Service Manual, Rabbit/Scirocco. 1975, 1976, 1977. Cambridge: Robert Bentley, 1977.